The **BIG** Resource Book

T0359237

THE PACIFIC

Ruth Naumann

NELSON
A Cengage Company

The Big Resource Book of the Pacific
1st Edition
Ruth Naumann

Text design and maps: Helen Andrewes
Cover design and illustration: Brenda Cantell
Reprint: Jess Lovell

Any URLs contained in this publication were checked for currency during
the production process. Note, however, that the publisher cannot vouch
for the ongoing currency of URLs.

Acknowledgements
The author wishes to acknowledge the following people and organisations
for their assistance in providing permission to use resources in this book:
Alexander Turnbull Library, National Library of New Zealand, Te Puna
Matauranga o Aotearoa for the photograph of kilikiti on page 64
Dominion Post Collection (PAColl-7327) and the following cartoons: page
21, (1) Heath, Eric, 1923- :[Jacques Cousteau]. "Here vee are at zee scene of
zee tests - zere eez no danger of nuclear nastiness - I sweem on gracefully,
because zee camera is following me". [1980s]. Reference No. B-143-074;
(2) Bromhead, Peter, 1933- :The South Pacific. 1 June 1981. Reference No. A
-331-130; (3) Lodge, Nevile Sidney, 1918-1989:'Pardon, Monsieur The
President, but as he was leaving, the New Zealand Deputy Prime Minister
scribbled on your front door!' Evening Post. 1973. Reference No. B-134-507
(4) Nisbet, Alistair, fl 1990s :Mururoa. 4 May 1992. Reference number: J
-023-001; (5) Heath, Eric Walmsley 1923- :"Good grief! Is it that time
again!". 1974. Reference No. C-132-125; (6) Heath, Eric Walmsley, 1923- :
Quickly, M'sieur! While no one's looking!! [23 August 1979]. Reference
No. B-144-003; page 30, Scott, Thomas 1947- :One day Mrs Speight will
regret this... Evening Post, 2 June 2000. Reference No. H-618-025; page
47, Scott, Thomas 1947- :Gee, those scientists are right about the oceans
hotting up ... - half of these fish are already cooked!!! [9 May 1989].
Reference No. H-196-011.
Michael Field for the photographs on pages 8, 11 (woman with mobile
phone), 53, 55; Greenpeace Aotearoa New Zealand for the photographs
on pages 35 whaling protest Greenpeace / Phil Crawford; 41 deep sea
protest Greenpeace/Roger Grace; 42 whaling Greenpeace/Sutton-
Hibbert; 43 whaling Greenpeace/Sutton-Hibbert; 44 Greenpeace; 45
Greenpeace/Dean Purcell; Vivienne Ross for the photographs on pages 11
village housing, 60 market; New Zealand Defence Force for the
photographs on page 32; New Zealand Ministry of Foreign Affairs and
Trade / NZAID New Zealand's International Aid & Development Agency
for the photographs on pages 9, 10, 33, 36, 37, 46; National Aeronautics
and Space Administration for photographs on pages 5 planet Earth, 16
tropical cyclone, 18 Cyclone Heta, 46 El Nino, 51 Kiribati; National Oceanic
and Atmospheric Administration (NOAA) for photographs on pages 19
and 26 Molokai Island

National Library of New Zealand Cataloguing-in-Publication Data
Naumann, Ruth.
The big resource book of the Pacific : social studies literacy / Ruth
Naumann.
ISBN 978 0 17 096269 8
1. Oceania—Social life and customs—Juvenile literature.
2. Oceania—Social life and customs—Problems, exercises, etc.
[1. Oceania—Social life and customs. 2. Oceania—Social life
and customs—Problems, exercises, etc.] I. Title.
306.0996—dc 22

Cengage Learning Australia
Level 7, 80 Dorcas Street
South Melbourne, Victoria Australia 3205

Cengage Learning New Zealand
Unit 4B Rosedale Office Park
331 Rosedale Road, Albany, North Shore 0632, NZ

For learning solutions, visit **cengage.co.nz**

Printed in Malaysia by Papercraft
16 17 18 24

Contents

The Pacific is Huge

RESOURCE 1   How the Pacific Got Its Name

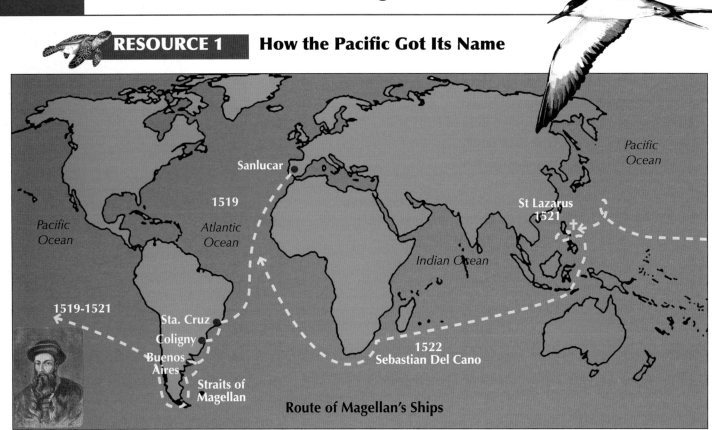

Route of Magellan's Ships

Ferdinand Magellan was the first European to sail across the Pacific. He came from Portugal although he sailed for Spain. He made the first known circumnavigation of the world. In 1520 Ferdinand's 3 wooden ships left the Atlantic Ocean for a new ocean. After the stormy Atlantic, this new ocean was gentle. Ferdinand called it Mare Pacificum. This is Latin for peaceful sea. It is where the name Pacific comes from. Ferdinand thought he would cross the Pacific in 2 or 3 days. It took him about 4 months. His crew got exhausted. They were starving. They ate sawdust, leather strips from sails, rats. Drinking water was stinking, slimy, and yellow. Scurvy caused swollen and bleeding gums, and blue and red marks on skin. Finally they got fresh fruit, vegetables and water at Guam.

1 Name the place in Spain that Ferdinand set sail from.

2 Suggest a reason that Ferdinand had to crew his ships with prisoners freed from prison in return for sailing with him.

3 Decide which word on the map is most likely to mean a narrow strip of water between land.

4 Make a comment about the feature on the map named after Ferdinand.

5 Explain how close to Australia the expedition got.

6 Account for the fact that Guam cured the crew's scurvy.

7 Ferdinand was killed by natives in the Philippines. Work out if he was still alive when the expedition sailed around the southern tip of Africa.

8 Give details about the name Pacific.

9 Explain the reason for Ferdinand's voyage.

10 List ways Ferdinand's voyage affected the sailors and would probably have affected the future of Pacific people.

RESOURCE 2 — What We Know About the Pacific But Magellan Did Not Know

- Largest of world's 5 oceans (followed by Atlantic, Indian, Southern, Arctic).
- In 2000 a 5th ocean (Southern Ocean = southern part of Pacific) was marked out; this took away part of Pacific.
- Includes many seas such as Tasman.
- Is about 15 times the size of US, covers about 28% of Earth's surface, is larger than total land area of world.
- Lowest point is Challenger Deep in Mariana Trench – nearly 11,000m.
- Has about 25,000 islands.
- Its people speak many languages e.g. English in Australia, NZ, Papua New Guinea (which has hundreds of languages), Solomons, Vanuatu, Fiji; French in New Caledonia and French Polynesia; Pidgin in PNG and Vanuatu.
- Is not always peaceful e.g. tropical cyclones, earthquakes, tsunamis.
- Has oil and gas fields e.g. in shallow water off coasts of Australia and NZ; metals, sand and gravel, minerals e.g. copper and gold in Papua New Guinea; fish e.g. tuna.
- Has environmental issues e.g. endangered species such as dugong, effects of nuclear tests.

1　Put all the information in this resource into a full page diagram. Make sure you explain the meaning of Pidgin and the location of the new Southern Ocean.

2　In your group create a star diagram of why the Pacific is important to your country. Share with the class. Add to your diagram.

1　Ferdinand Magellan's parents had died when he was 10. Find another 10 facts about him.

2　Other early European explorers to see the Pacific were Vasco Nunez de Balboa from Spain (before Magellan), Francis Drake and Captain James Cook from England (both after Magellan). Find out when they came.

3　Find out the sizes of the 5 oceans of the world.

4　Find 10 features that mark the Pacific as being different to other oceans.

Distribution of Islands

 **RESOURCE** **Location of Pacific Isalnds**

1 Check out the map. Note which islands you have heard of and which ones you have not heard of. Note how their names are spelled. Then turn your book over so you can not see the map. List as many islands as you can. Do this several times until you get familiar with the names and locations.

2 Write a few sentences about the location of your country in relation to the Pacific and its other places.

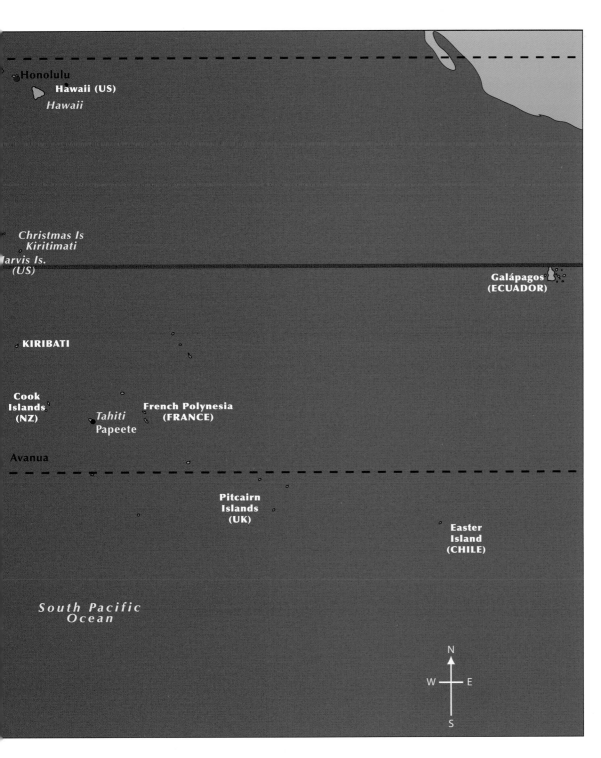

The aim is for everyone in the class to have a short profile on every Pacific country. Decide how you will get this outcome. Decide time allowed for research, and headings such as how their names are pronounced (eg Kiribati is said as Kiribass). Decide how you will present your research to the class eg handout, PowerPoint. Do your research and presentation to the class.

Location

| RESOURCE 1 | **Regions of the Pacifc** |

Europeans were excited that their explorers had sailed into the Pacific. A new ocean, new islands and new people had arrived on the scene. Two Frenchmen used Greek words to describe regions of this new place.

nesos = islands
poly = many
melas = black
mikros = small

→

region = an area of Earth's surface which contains features that make it different from other areas
location = where things are found on Earth
distance = space between different locations on Earth
distribution = arrangement of things on Earth
movement = change in location of things across Earth

Region	Melanesia	Micronesia	Polynesia
Meaning	black islands	small islands	many islands
Location	south of Equator	mostly north of Equator	island groups big distances apart
Size of islands	large	thousands of small islands	small coral islands and atolls mixed with volcanoes
Indigenous (native) people	generally shorter, darker skin, black woolly hair	generally taller, lighter skin, wavy hair	generally tallest, golden skin, straight to wavy hair
Example of islands	Papua New Guinea, Solomons, Vanuatu, Fiji, New Caledonia	Fed States of Micronesia, Palau, Marshall Is, Nauru, Guam, Northern Mariana Is, Caroline Is	Kiribati, Niue, Cooks, French Polynesia, Tonga, Samoas, Tuvalu, Tokelau, Pitcairn

(There are other islands in the Pacific such as those close to North and South America eg the Aleutian Islands [USA]. They are located outside the regions of Melanesia, Micronesia and Polynesia.)

1   Explain how the 3 regions got their names.

2   Give 3 general differences among the 3 regions.

3   Write a sentence about the distribution of islands in Polynesia.

4   Give a reason that you will never see lines marking out Polynesia, Micronesia and Melanesia if you sail around the Pacific.

5   Polynesia is said to be shaped like a triangle. Invent a way to test this.

6   Fiji's culture is largely Polynesian. Explain why that is unexpected.

Papua New Guinea (Melanesia).

Caroline Islands (Micronesia).

Tokelau (Polynesia).

RESOURCE 2 Where do New Zealand and Australia Fit In?

Oceania describes the region between Asia and the Americas. The Australian continent is the major landmass. Also included are the thousands of Pacific Islands. New Zealand is one (or 3).

Oceania is sometimes divided into Melanesia, Micronesia, Polynesia and Australasia. In this definition, Australasia means Australia and New Zealand.

Some definitions say the Pacific Islands mean Oceania without Australia.

Most definitions put New Zealand as part of the Polynesian triangle. The Maori are a major culture of Polynesia.

New Zealand and Australia have indigenous people – Maori and Aborigine. But the majority of their people have come from Europe more recently. In New Zealand and Australia, the term Pacific Islander does not usually refer to those people of European origin. It refers to people from other Pacific islands such as Tonga.

And just to make it even more complicated, Earth is divided into 8 ecozones. An ecozone is a major ecological region (plants and animals and their environments). Oceania is one of the ecozones. It includes all Micronesia, Fiji, and all Polynesia except New Zealand. New Zealand and Australia, along with New Guinea, Solomon Islands, Vanuatu and New Caledonia make up the separate Australasia ecozone.

No matter what the terms mean, the Pacific Ocean laps coasts of your country. And your country is very involved in the Pacific.

---

1 Find the word, used 3 times in the resource, that means big.	4 Suggest a reason why, if you look up meanings of terms like Melanesia, Oceania and Australasia on the net, you will get many different answers.
2 Name the two indigenous people of NZ and Australia.	
3 Explain what an ecozone is.	

RESOURCE 3 The Terms Pacific Rim and Pacific Basin

Rim refers to those countries bordering the Pacific Ocean such as Peru. Basin includes all the island countries.

Draw a soup bowl. Label it to show the part called 'rim' and the part called 'basin'.

Anthropology is the study of human societies. An anthropologist is a person who studies human societies. Anthropologists would love to know where the people in the Pacific come from.

Because anthropologists do not know for sure, they make up theories. A theory is an educated guess. It is based on as much evidence as can be found. One theory is that the Pacific people came from Southeast Asia.

Thor Heyerdahl (1914-2002) was a famous marine biologist from Norway. His theory was that many people in the Pacific came from South America, not Asia. He thought Stone Age humans sailed from Peru to the Polynesian islands on big rafts built from balsa logs. Thor wanted to prove his theory. He and a small team went to South America. They built a raft of balsawood. It had a primitive sail. They called it Kon-Tiki.

In 1947 Thor and his crew set sail from Peru. The only modern technology they had was a radio, military rations, and fresh water in small cans. They sailed across the Pacific. Their voyage ended when the raft hit a reef off the Tuamotu Islands near Tahiti. It took 101 days to sail 8000 km.

1 Comment about whether or not you would be interested in being an anthropologist.

2 Give a piece of evidence to show that Thor was interested in anthropology.

3 Outline Thor's theory and how he tried to prove it.

4 Explain the difference between Thor's theory and another theory.

5 Decide why anthropologists do not know for sure where the Pacific people came from.

USE THE NET

1 Find out 3 facts about balsawood.
2 Find out why the raft was named Kon-Tiki.
3 Find out another 10 facts about Thor Heyerdahl.
4 In 2006, a grandson of Thor Heyerdahl was one of a 6-member crew that set sail to repeat Thor's voyage. Find out how it went.
5 Find out what the most popular present theory is about where the Pacific Island people came from.

UNIT 4 Names of Islands

RESOURCE 1   Nations, Dependencies and Territories

Some places rule themselves. They have their own governments. They are called nations.

Some places do not completely rule themselves. They depend on a nation or are part of the territory of a nation. So they are called dependencies or territories.

Nations:	Dependencies and territories:	
Australia	American Samoa (USA)	Midway Islands (USA)
Fiji	Baker and Howland Islands (USA)	New Caledonia (France)
Kiribati		Niue (NZ)
Marshall Islands	Cook Islands (NZ)	Norfolk Island (Australia)
Micronesia	Coral Sea Islands (Australia)	Northern Mariana Islands (USA)
Nauru	Easter Island (Chile)	
New Zealand	French Polynesia (France)	Palmyra Atoll (USA)
Palau	Galapagos (Ecuador)	Pitcairn Islands (United Kingdom)
Papua New Guinea	Guam (USA)	
Samoa	Hawaii (USA)	Tokelau (NZ)
Solomon Islands	Jarvis Island (USA)	Wallis and Futuna Islands (France)
Tonga	Johnston Atoll (USA)	
Tuvalu	Kingman Reef (USA)	Wake Island (USA)
Vanuatu		

RESOURCE 2   Photographs

Make up 5 questions from the resources. For example: Are there any cities in the Pacific outside NZ and Australia? Find answers on the net.

The Pacific Ring of Fire

RESOURCE 1 **Plate Tectonics**

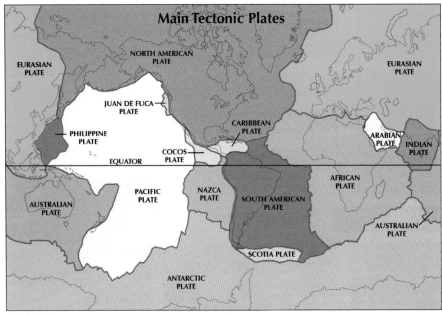

Scientists have developed a theory called plate tectonics. Tectonics comes from a Greek word meaning carpenter. Tectonics is about how the Earth's crust is built and what forces it to move. The crust is the surface of Earth. It is covered by land and oceans. The plate tectonics theory says the crust is made of plates. There are 7 big plates and about 20 smaller ones. They can be 5 to 100 km thick. Plates move at speeds of about 2.5 cm a year. That is about as fast as your fingernails grow.

Where two plates meet is known as a boundary. If plates collide, one may go under the other. This is called subduction. It causes big trenches. The Mariana Trench is formed by the Pacific Plate sliding under the Philippine Plate. Challenger Deep is there.

1 Decide if the plates lie under the ocean and continents. Offer evidence to back up your decision.

2 Find two differences between the Pacific Plate and the Australian Plate.

3 Name the plates which form boundaries with the Pacific Plate.

4 Name the plates on which NZ is located.

5 Name the plates on which Australia is located.

6 Give the name for the lines between the Pacific Plate and the other plates such as the Australian Plate that border it.

7 Draw a diagram to show what 'subduction' means.

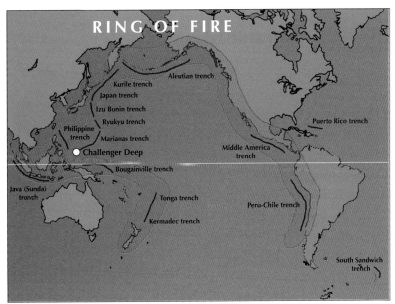

RING OF FIRE

The Ring of Fire is a name given to a zone of frequent earthquakes and volcanic eruptions. The zone partly encircles the Pacific. It is called a ring but it is shaped more like a horseshoe. It is about 40,000 km long. The earthquakes and volcanic eruptions are caused by the movement and collisions of tectonic plates.

Most of the world's major volcanoes are here. A volcano offshore from Papua New Guinea is called Blup Blup.

Most of the world's earthquakes also happen here. In 2006 an earthquake near Tonga measured 8.1 on the Richter scale. It set off a false alarm about a big tsunami heading for places like New Zealand.

The Pacific Tsunami Museum is in Hilo, Hawaii. Tsunamis have killed more people in Hawaii than all other natural disasters put together. The Museum is to teach people about tsunamis so no one will ever again die in one. It is also a memorial to people killed in past tsunamis.

1   Name the big tectonic plate in the Ring of Fire.

2   In no more than 100 words, describe the location of the Ring of Fire.

3   Most volcanoes form along the edges of Earth's tectonic plates. Use the Ring of Fire as evidence for this statement.

4   Name 3 natural disasters that people in the Pacific might face.

Find 12 pieces of advice on how to be safe during an earthquake, a volcanic eruption and a tsunami (4 pieces of advice for each natural disaster).

Types of Islands

RESOURCE 1   Volcanic Islands

Some Pacific islands are volcanic. They have been built by volcanoes. Magma (molten rock) was forced through vents called hot spots. Or by plates overriding one another at boundaries. Examples are Hawaii, Guam, the main islands of Fiji. A new volcanic island emerged from the sea in Tonga in 1995.

Some Pacific islands are atolls (see RESOURCE 2). Baker and Howland are uninhabited atolls. Johnston Atoll is a coral atoll of 2 islets. Palmyra Atoll is an atoll of about 50 islets. Tokelau is 3 atolls. Niue is one of world's biggest coral atolls.

Some Pacific islands are a mix of volcanic islands and atolls. Cook Islands are coral atolls in the north, hilly volcanic islands in the south. Pitcairn has a volcanic island, 2 coral atolls, a sandbar and an uplifted coral island. America Samoa is 5 volcanic islands and 2 atolls.

A Pacific volcanic island.

1 Name 2 countries made of volcanic islands, 2 countries made of atolls, 2 countries made of a mix.

2 Decide if volcanic islands rise from the ocean floor or not. Account for your decision.

RESOURCE 2   Atolls

Reef building is a biological process. Atolls are an example of how biological processes can make changes to Earth's surface.

Charles Darwin (see page 27) created a theory about how atolls are formed. Today many experts still accept his theory. He said an atoll begins when a volcano rises out of the ocean fringed by a coral reef. Wind and sea wear the top of this island down. The top begins to sink. With it goes the fringing reef. Reef animals can only survive near the surface of water. New reef animals build on top of the limestone skeletons of old reef animals. A barrier reef forms. It gathers soil and seeds such as coconut palms. The sinking island disappears under water. It leaves a circle of land made up of low coral islets, with a shallow lagoon in the middle.

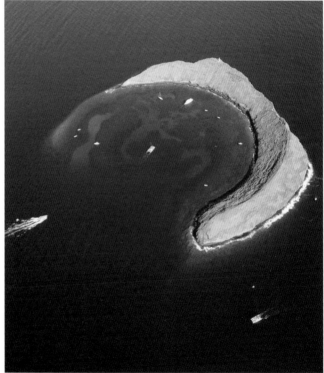

A Pacific atoll.

Create a series of diagrams to show how an atoll is made. Label your diagrams.

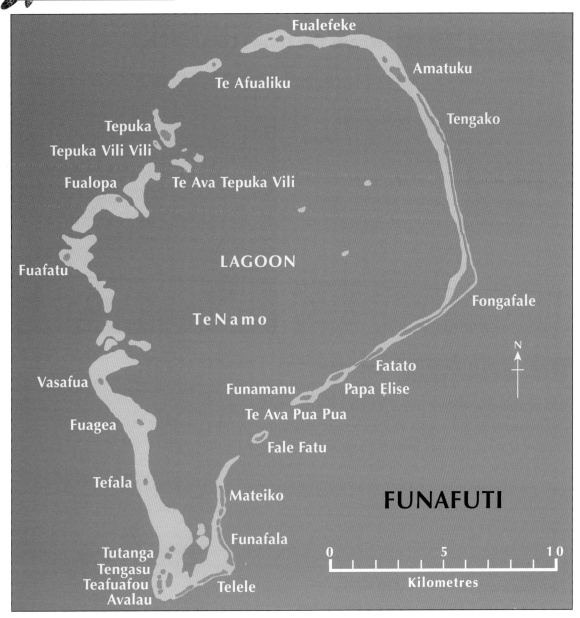

Tuvalu is made up of 9 coral atolls. Funafuti is the chief atoll in Tuvalu. It is made up of islets. It is 21 km long. The total land area is about 2.6 square km. An expedition in 1897 proved Darwin's theory of atoll forming by sinking a 335m bore into the island.

1   Make your own copy of the diagram. Give it a key / legend. Colour it.

2   Explain which part is the old volcano top.

3   Explain which part is the old fringing reef.

1   Atolls are some of the most complex structures on Earth. Scientists have been studying them for years but many mysteries remain. Find out some ideas about atolls.

2   Write down 5 questions you would need answering for you to become an expert on Pacific volcanic islands. Find the answers. Share with the class.

Tropical Cyclone

RESOURCE 1   Where the Terms Come From

A tropical cyclone is also called a typhoon and a hurricane.

= typhoon (Asia, North Pacific; from Greek *tuphon* [whirlwind] and Chinese tai fung [big wind])

= hurricane (USA; from Spanish *huracan*)

= cyclone (NZ, Australia, South Pacific; from Greek *kyklon* [moving in a circle])

List conclusions you can come to from seeing where the words come from.

RESOURCE 2   Cyclone

A cyclone is a violent storm. Tropical means the cyclone comes from between the tropics of Capricorn and Cancer. A tropical cyclone always comes from out at sea. It is 'T' on a weather map.

A cyclone moves in a circle. In the Northern Hemisphere (above the Equator), it moves in an anticlockwise direction. In the Southern Hemisphere (below the Equator) it moves in a clockwise direction. The centre of the cyclone is called the eye. It is the calmest part. Cyclone season in the Pacific is between November and April. On average there are 8–10 cyclones per season. Each lasts between a few hours to 2 weeks. The average is 6 days.

The route a cyclone takes is called a track. It moves at about 15-25 km/h. But it might suddenly take off in another direction. It might stop for a bit. If it travels over cooler seas, such as those south of Capricorn or over big land masses such as Australia, it will fade. However, it may come to life again.

In April 1968, Cyclone Gisele came to life again as it moved south over NZ. Winds gusted to 270 km/h in Wellington. They sank the inter-island ferry *Wahine*. 51 people died. Cyclone Tracy hit Darwin on Christmas Day 1974. It had winds up to 217 km/h. It killed 65 people.

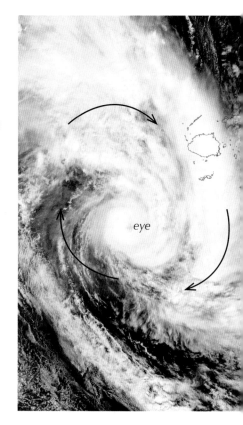

eye

1 Explain these terms:
   a) tropical cyclone,
   b) Northern and Southern Hemispheres,
   c) eye of the cyclone,
   d) track of the cyclone
   e) Capricorn and Cancer.

2 Describe the weather in your country at the time of the cyclone season.

3 Explain the difference between tropical cyclones in the 2 hemispheres.

4 Choose one word to describe a cyclone. Explain your choice.

RESOURCE 3   The Pacific is a Cyclone Region

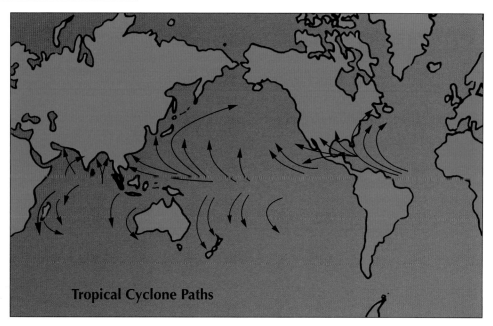

Tropical Cyclone Paths

1 Compare the Pacific with other regions of the world in terms of cyclones.

2 Suggest a reason that not all islands of the Pacific are marked on the map.

3 Decide if you would put the danger to your country for cyclones as High Risk, Moderate Risk or Low Risk. Give a reason for your choice.

4 Select the fact from RESOURCE 2 that the map best illustrates.

5 Create a key / legend for the map.

RESOURCE 4   Measuring Tropical Cyclones

Different regions use different scales to measure tropical cyclones. Australia and New Zealand use the following Saffir-Simpson scale.

Category One:	Category Two:	Category Three:	Category Four:	Category Five:
Winds 119-153 km/h. Storm surge generally 1-2m above normal. Eg coastal road flooding and minor wharf damage.	Winds 154-177 km/h. Storm surge generally 2-3m above normal. Eg campervans and billboards flip.	Winds 178-209 km/h. Storm surge generally 3-4m above normal. Eg large trees blown down.	Winds 210-249 km/h. Storm surge generally 4-6m above normal. Eg power poles snap.	Winds over 249 km/h. Storm surge generally greater than 6m above normal. Eg trains overturned, heavy cars lifted and thrown.

1 The Saffir-Simpson scale has a lot more detail than the summary here. Explain what a summary is and why it is useful.

2 Describe the 3 categories chosen for this summary.

USE THE NET

1 Find out how the Saffir-Simpson scale got its name.

2 Find more examples of tropical cyclones that have hit the Pacific.

3 Find out if technology can help reduce cyclone risk.

Cyclone Heta

 RESOURCE 1 **Cyclone Heta's Track**

Niue is a self-governing state in free association with NZ which administers Niue's foreign affairs and defence policy. Cyclone Heta hit Niue in January 2004 on a Tuesday.

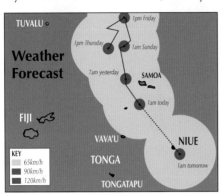

Satellite image of Cyclone Heta.

1 Did the cyclone come in, or did it come out of, the cyclone season?

2 Describe the force of winds closest to the eye of a cyclone.

3 Which country was right in the middle of the projected path of Heta?

4 Can you work out a meaning for the word 'projected' as used in the previous question?

5 At which point does the track become a projection?

6 On Saturday, Tokelau was hit with strong winds and thunderstorms. What does the graphic suggest about the location of Tokelau?

7 On Tuesday, gale-force winds hit Tonga but it missed the worst of the cyclone. How accurate was the projection for Tonga?

8 What day is referred to as 'yesterday'?

9 What day is referred to as 'today'?

10 The winds that hit Niue had speeds of 230 km/h and gusts reaching 275 km/h. How accurate was the projection for Niue?

11 In which country did the cyclone most likely destroy crops, block roads and knock out bridges on Monday?

RESOURCE 2 **Cyclone Strike**

Radio and television warned the 1300 to 2000 people of Niue that Heta was on its way.

To protect it, officials took down the satellite dish that provided phone contact. People bought candles and emergency supplies. They bunkered down. Some who lived near the sea went inland to camp in empty houses. They listened to officials advising them to protect crops by cutting leaves off banana plants to leave only stems. But they could not protect other plants such as breadfruit, pawpaws and grapefruit. They worried about avocado crops due to be harvested soon.

Cyclone Heta struck. Its winds whipped the sea into 50m waves. Destroyed houses washed up to 100m inland. Heta smashed coastal caves. A wave knocked a Filipino voluntary worker out of a house, over the next house and behind a water tank. Other people were injured. Vehicles were ruined. A car was tossed into a tree. Waves raced up 30m cliffs. People fled through the bush.

A nurse called Catherine shielded her baby son Daniel as the cyclone wrecked her home. Catherine was crushed to death. Niue's hospital was flattened and medical

Niue after Cyclone Heta struck.

supplies washed away. The island's two doctors set up a makeshift hospital in the Public Works Department.

Much of the town of Alofi, the capital, was flattened. The satellite dish was damaged. Communication and electricity were cut. Water was in short supply. Many septic tanks were washed away. Petrol was rationed. Although the airport stayed open, many roads were closed. People helped themselves to surviving goods in stores. Some looted wrecked houses. Catherine's mother started the hunt for her daughter and grandson. Catherine was found under rocks and mud. Daniel was still alive. He was flown to the Starship Hospital in Auckland.

NZ and Australia sent planes with emergency supplies. Relief operations were launched to get money for the people on Niue. The Premier of Niue called on the 20,000 Niueans in NZ to come home and help rebuild. Communities held prayer meetings.

The clean-up began. Stranded tourists helped. The cyclone released particles of deadly asbestos into the air from houses built by NZ in the 1960s and 70s. Locals added to it by burning wreckage of roofs and walls containing asbestos. Hungry wasps, bats, rats, and bees came out of the bush looking for food and water. The smell of raw sewage started to drift. Stories went round such as how one man survived by hanging on to jungle vine.

Experts worked out how many millions would be needed to rebuild Niue. Some people spoke about the death of Niue. They said its population would fall below 500 and there would not be enough people to keep Niue going.

---

1 Work out why Niue had many empty houses to offer inland shelter.

2 List the physical problems that people faced after the cyclone hit.

3 List other, unseen problems, the cyclone might have caused.

4 Make a report on what immediate help was supplied after the cyclone.

5 Make a report for what long-term help would probably be needed.

6 Decide what might have been your attitude to the cyclone if you were a tourist operator on Niue.

7 Find the terms for
  a) leading government minister,
  b) dangerous mineral used in building,
  c) areas where people live,
  d) people from Niue,
  e) to get goods illegally.

1 Find out how cyclones get their names, who names them, and if there are any 'male' cyclones.

2 Find out what the situation is in Niue today and how well it survived Cyclone Heta.

An Historical Issue – Nuclear Testing in the Pacific

RESOURCE 1 — An Historical Issue – Nuclear Testing in the Pacific

Where are Bikini and Enewatak Atolls?

What is radiation?

What is Greenpeace?

What happened to the French secret agents?

Where does the International Court of Justice sit?

What is the other name for Kiritimati?

Where had the French tested before Mururoa?

Why did they have to stop testing there?

What happened to the Rainbow Warrior?

What is an injunction?

About 240 nuclear tests were done in the Pacific by 3 western countries between 1945-1996. The USA tested on Johnston Atoll, Bikini Atoll, and Enewatak Atoll. The cleanup to remove radiation is still going on. The British tested on Kiritimati Island — and in Australia.

In the 1960s the USA and Britain stopped testing in the Pacific. But the French began testing in Mururoa, an atoll in French Polynesia.

Many citizens in the Pacific, including groups in Australia and NZ, protested against the tests. They said if the tests were as safe as France said, why did the French not do them in France? Vessels sailed to Mururoa to protest.

In 1985 French secret agents exploded the *Rainbow Warrior*, the Greenpeace ship, in Auckland. It was getting ready to lead an anti-nuclear protest fleet to Mururoa. The explosion killed a photographer on board.

In 1973 NZ and Australia took France to the International Court of Justice. The Court put an injunction on tests. France ignored the Court and all the protests. It put all its tests underground. The Court said France had done enough. It refused to act further.

1 Work with a partner or group to find out the answers to the questions.

2 Rewrite the resource including all the facts that you have found out.

RESOURCE 2 — Cartoons 1, 2, 3, 4, 5, 6

1 Match the summaries below to the cartoons on page 21.
a) French President Pompidou versus Kiwi tagger
b) Design label fails to appeal
c) President Pompidou casts a shadow
d) Famous explorer Jacques Cousteau does his bit for France
e) NIMBY (Not In My Back Yard) dumpers
f) Hidden damage

2 Choose ONE cartoon to write about. Include setting, people, actions of people, feelings of people, how cartoonist has identified people, message of cartoon.

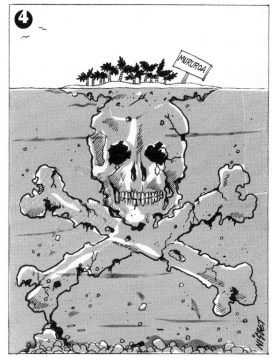

Quickly, M'sieur! While no one's looking!!

USE THE NET

1   Find out what the Doomsday clock is and where it is.
2   Find out what the present nuclear policy is in the Pacific.

UNIT 10 Mysteries of Easter Island

RESOURCE 1   People

Easter Island is often called the most isolated inhabited island on the planet.

Its Polynesian name is Rapa Nui. Its people and culture are known as Rapanui.

When Europeans arrived in the 1770s, they found 2 rival groups. They were Hanau Momoko and Hanau Eepe. Those were translated wrongly to mean Short Ears and Long Ears. This probably came from early visitors seeing some Rapanui had elongated earlobes into which they put wedges of wood. Hanau Eepe means fat. Hanau Momoko means thin. Maybe that was how Rapanui described their haves and have-nots.

In the 19th century slave traders kidnapped some Rapanui. They made them work for rich people in Peru. Missionaries came to Rapa Nui. So did diseases such as smallpox.

1 Niu is the Polynesian word for big. There is another island called Rapa, 650km south of Tahiti. Decide what size it will be in comparison to Rapa Nui.	2 Give one word to describe the relationship between Hanau Momoko and Hanau Eepe. 3 Draw a circle diagram to show possible results on Rapa Nui of the slave traders.

RESOURCE 2   Map

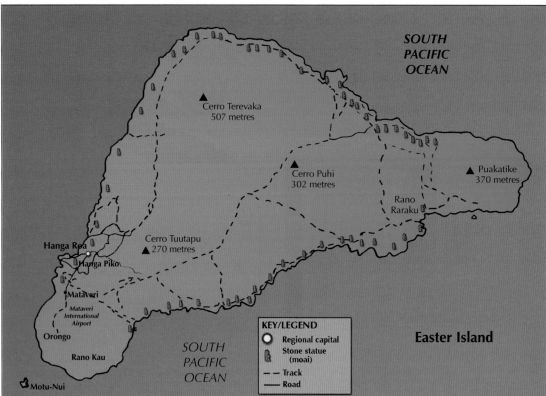

1 Name the a) capital, b) airport, c) highest point. 2 Describe the location of those 3 features. 3 Describe the shape of the island. 4 Explain where the moai are situated.	5 Present 2 pieces of evidence that proves Hanga Roa is an important place. 6 Decide if Rapa Nui is an atoll or if it is a volcanic island. 7 Suggest why Rapa Nui is a tourist attraction today.

RESOURCE 3   Moai

The biggest archaeological site in the Pacific is on Easter Island. It is a World Heritage site. It has hundreds of moai — giant stone statues.

Moai are aringa ora – living faces. They used to rest on long platforms called ahu. They looked inland. Perhaps they were to protect the villagers.

There are many theories about who built the moai. Among the more far-fetched is that extra terrestrials brought them. Another is that marooned aliens planted them as signals to fellow aliens to rescue them. Another is that an ancient society with the capability of flight built them. Another says the island was a rest and recreation station for space travellers who used blow torches and thermal lances to carve the moai.

Most moai were made in the Rano Raraku quarry. One theory is that workers walked the moai to ahu by using ropes. Another is that workers pulled the moai on sledges or log rollers. Another says the moai were carved around the quarry so that when the volcano erupted, it blew the moai to their ahu. Another theory says typhoons scattered ships that carried war elephants; some elephants arrived at Rapa Nui and they shifted the moai. By 1840 all the moai had been thrown off their ahu. What caused this? Earthquake? War?

1. Explain what the following do: typhoons, space travellers, archaeologists, quarry workers.

2. Imagine you are a moai carver. Describe your work and its dangers.

3. Think of possible solutions to the mysteries of how people moved the moai, and why moai moved off their ahu.

4. Share your thoughts with your group.

5. Decide on the group's best solution for each mystery and present them to the class.

RESOURCE 4   Rongorongo

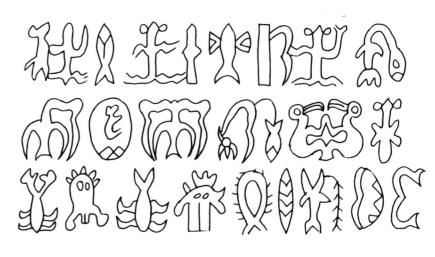

Researchers found wooden tablets on the island. They have 'script' on them which has become known as rongorongo. No expert has been able to deciper it.

Theorists who like extra-terrestrial explanations say the island was an energy source. Power surged up through the moai and out their eyes as laser beams. These laser beams carved the rongorongo script.

1. Describe what rongorongo looks like to you (eg writing? pictures?).

2. Draw another 3 symbols that might have been rongorongo.

3. Discuss with your group some theories about what rongorongo was.

4. Share your theories with the class.

RESOURCE 5   Trees

A palm forest used to cover Rapa Nui. Recent research suggests there may have been 16 million palm trees. Yet they were all destroyed. Did an environmental disaster kill them? Evidence based on pollen analysis supports the theory that the people cut down the trees.

The climate is cooler than most other Polynesian islands. Breadfruit, coconut palms and kava did not grow. A recent theory says islanders may have used palm sap as a food.

They used wood for fuel. They used it also to build canoes in the early days after arrival. But their seafaring technology got worse rather than better. Bones in middens (rubbish heaps) show the islanders ate a lot of dolphin at first. But then they began to eat food sources closer and closer to land.

No trees means no building of canoes suitable for long sea voyages. No canoes means no escape. Captain Cook in 1774 reported the islanders had poor canoes and no wood. Why would the people let themselves be trapped like this? Or did they think all other lands had sunk beneath the sea and that they were totally alone in the world?

1   Explain the link between eating dolphins and seafaring technology.

2   In your group talk about possible reasons for the trees disappearing.

RESOURCE 6   Birdman

The Rapanui held a Birdman competition each September. Each clan chief chose a person to be his representative. The chosen set off on reed rafts for Motu Nui. There they waited in a cave for terns. These seabirds came each year to lay eggs at the top of a cliff. The chosen climbed up and tried to snatch a freshly-laid egg. Then they had to swim back through the shark-infested ocean to Orongo. The first swimmer to hand his chief an unbroken egg was honoured. Most honour went to the chief who was named Tangata Manu (Birdman) for the coming year. His clan got special privileges. The last Birdman competition was in the 1860s.

Was there a special link between Rapanui and birds? Pacific Islanders saw seabirds as signs, messengers, spirits of the gods, the dead. But on Rapa Nui even the moai have wing-like hands. Were the Rapanui sad that they did not have the option of leaving and so they admired birds who could leave?

1   Explain what these are: reed raft, clan, tern, honour, link, option, shark-infested.

2   Imagine your chief has chosen you for the Birdman competition. Describe your actions and feelings as you go through the competition.

USE THE NET

1   Find out how Easter Island got its name.

2   Rabbits were introduced in the 1860s. They became extinct. This may be the only place in the world where this happened. Find out why it happened.

3   Scientists use several dating methods on Easter Island today. Describe how one method works.

4   Chile took over the island in 1888 but it was not until 1966 that Rapanui got citizenship. Describe the relationship between Chile and the Rapanui today, especially over the issue of land.

Historical Places in the Pacific

 RESOURCE 1   **Tinian Bomb Launch**

Tinian is part of the Commonwealth of the Northern Mariana Islands. From Tinian the Americans launched the 2 atomic bombs they dropped on Japan in 1945. They dug 2 pits. They lowered the bombs into the pits. They drove the planes over the pits. The planes' bellies opened. The bombs were hydraulically lifted inside. Locals say the foliage is strange near the pits today. For example, coconut trees grow fruit that does not ripen.

RESOURCE 2   **Pearl Harbour**

The Japanese attacked this US base in Hawaii in 1941. World War 2 had been going since 1939 but the US was not fighting in it. The attack damaged and destroyed US warships and planes. It killed and wounded US servicemen and civilians. The next day the US declared war on Japan.

Pearl Harbour memorial.

RESOURCE 3   **Norfolk Island Convict Settlement**

This island off Australia got a reputation for being a place of cruelty. A bishop reported how men condemned to death went down on their knees to thank God. Spared men cried. By 1855 public pressure closed the convict settlement.

RESOURCE 4   **Bounty's Landing on Pitcairn Island**

In 1789 the British Royal Navy sent Lieutenant William Bligh, commander of the ship *Bounty*, to Tahiti. Fletcher Christian, a crewman, led a mutiny. He ordered Bligh, with 18 others, into the *Bounty's* launch. Bligh navigated the overcrowded open launch on a long voyage to safety with only a sextant and pocket watch. The mutineers sailed to Pacific islands. They got females on board. When they found Pitcairn Island, they burned the *Bounty*. Some of their descendants still live on Pitcairn.

 RESOURCE 5   **Tahiti in Paintings**

Paul Gauguin was a poor artist from France. People at the time did not see his talent. In 1891 he sailed to Tahiti. He painted Polynesians. Still poor, he died on the Marquesas Islands in 1903. After his death, people realised his great work in developing modern art.

RESOURCE 6   **Home of Typee Cannibals in the Marquesas Islands**

Herman Melville was an American looking for adventure. He shipped out as a cabin boy on a whaler in 1841 for the Pacific. The captain was cruel. Herman deserted. Typee cannibals in the Marquesas Islands captured him. Another whaling ship rescued him. Herman wrote a book about his time with the Typee. But his most famous book was *Moby Dick*.

Herman Melville.

RESOURCE 7   Site of Cook's Death in Hawaii

Explorer James Cook was back in Hawaii in February 1779. Storms had damaged his ship. It needed repairing. The Hawaiians had been friendly before. But this time there was tension between sailors and islanders. James went ashore to check it out. A fight broke out. The islanders killed James.

RESOURCE 8   Ruins on Malden Island

Malden is in Kiribati. It is named for the British navigator who sighted it in 1825. Nobody was living on it. But it had ruined buildings. Who had built them? Wrecked seamen? Pirates? South American Incas? Early Chinese navigators? An archaeologist examined the ruins in 1924. He said a small Polynesian population had lived there some centuries earlier.

RESOURCE 9   Aggie Grey Hotel in Samoa

Aggie Grey hotel.

Aggie Grey's father was a chemist who had migrated to Samoa. Her mother was Samoan. In WW2, Aggie bought the site of a local hotel. It was a small wooden building but it got a reputation for helping American servicemen stationed in Samoa. As it got grander, it became a legend in the South Pacific. Hollywood stars and royalty visited. Aggie Grey is said to be the inspiration for James Michener's character Bloody Mary in his book *Tales of the South Pacific*.

RESOURCE 10   Guadalcanal Battlefields

Guadalcanal is an island in the Solomons. The Japanese captured it during WW2. The Allies and US troops fought to get it back from August 1942. The Japanese put heads of dead Americans on pikes and planted them around the Marine base. Malaria killed many soldiers on both sides. It was hard getting supplies through. The Japanese called the island 'Starvation Island'. They withdrew in February 1943.

US marines landing in Guadalcanal 1942.

RESOURCE 11   Leper colony on Hawaii's Molokai Island

Molokai Island leper colony.

Leprosy destroys nerves and skin. When it arrived in Hawaii in 1865, Kalaupapa on Molokai Island was made a leper colony. Sea surrounded it on 3 sides; the other side had high cliffs. The first lepers lived in caves or between rocks, or in huts of sticks and leaves. Sometimes they were told to jump off the ship bringing them to the island and swim for their lives. The crew threw supplies into the water. After 1969 no more lepers were sent there.

RESOURCE 12　Vailima in Samoa

Robert Louis Stevenson, the author of *Treasure Island*, and *The Strange Case of Dr Jekyll and Mr Hyde* went to Samoa in 1890. He had tuberculosis. The Pacific air might help him. The Samoans called him Tusitala — teller of tales. He died there 4 years later. Today his home, Vailima, is a museum.

Vailima.

RESOURCE 13　Galapagos

Charles Darwin was a British naturalist. He visited and explored Galapagos in 1835. His findings led to his theory of natural selection. The theory explained how all species on Earth had come from a simple, single-celled ancestor. It changed the way scientists thought about life.

1　Make a chart about reasons people had for coming to the Pacific.

2　Arrange the resources in order of interest to you.

3　Make a list of 5 – 10 questions you would like answered about any of the resources eg What were the names of the atom bombs loaded at Tinian? Why did the Navy send Bligh to Tahiti?

USE THE NET

Find out the answers to your questions.

Colonisation and Decolonisation

RESOURCE 1   Terms

Today, you live in an independent country. That means it rules itself. It was not always independent. At one stage it was a British colony. It was part of what was called the British Empire. Other places, like Canada, India, and Pacific countries were part of this Empire. And other big powers, such as France and Germany, also had colonies in the Pacific.

To get the colony, the big power might send its own people to run it. This is annexation. Or the big power might claim the place because it 'discovered' it. Or the League of Nations might give a colony to the big power. This is called getting a mandate. Or the big power might sign a treaty with the indigenous (native) people such as the Treaty of Waitangi between some NZ Maori chiefs and Britain. Making another country a colony is called colonisation.

A country might be colonised by more than one big power. The people of Palau, for example, have had Spanish, German, Japanese and US rulers.

A protectorate was a country protected and partly controlled by a big power. Even Tonga, which had its own kingdom, became a British protected state.

Territory meant the same. Another term was influence. One country was said to have influence in the other.

Term	Meaning
empire	group of countries ruled by one other ruler or country
colony	country ruled by another country
League of Nations	international organisation set up after WW1 to stop wars – it failed
annexation	comes from Latin word meaning 'tied to'
mandate	official instruction
colonisation	also known as empire-building
protectorate	think of the modern term protective custody

1   Make a list of terms from this resource, with their meanings, that will help you understand the unit.

2   The term colonisation is sometimes used in anger by people living in a former colony. In your group suggest reasons for this.

3   In your group talk about reasons for big powers wanting to have colonies.

RESOURCE 2   Examples of Colonisation

Germany — Papua New Guinea, Solomon Islands, Palau, Federated States of Micronesia, Caroline Islands, Nauru, Mariana Islands, Marshall Islands, Samoa

France — Vanuatu (with Britain), New Caledonia, French Polynesia, Wallis and Futuna

Britain — Australian colonies, New Zealand, Fiji, Nauru, Kiribati, Tuvalu, Vanuatu (with France), Pitcairn, Tokelau, British Solomon Islands, Samoa, Tonga, Papua New Guinea, Cook Islands

1   Name 3 places that appear on two lists. Suggest a reason for this.

2   Decide how you could use this resource as evidence of colonisation in the Pacific.

RESOURCE 3   Results of Colonisation on the Pacific islands

Name the results of colonisation shown in the pictures. In your group suggest other results. Share with the class.

RESOURCE 4   Decolonisation

Decolonisation means swapping rule by a big power for rule by the colony. It is violent when people demand self-government and fight to get it. It is peaceful when the big power prepares citizens to govern themselves, and then gives them independence.

Decolonisation came to the Pacific in the 20th century. In 1962 Western Samoa (it dropped Western from its name in 1997) became independent from NZ. It was the first Polynesian country to get its independence back. Other examples were Vanuatu in 1980, Tonga and Fiji in 1970, Solomons in 1978, Papua New Guinea in 1975.

Some Pacific islands chose to keep a close relationship with their big power. For example, the Cook Islands in 1965 and Niue in 1974 chose to have self-government in 'free association' with NZ. This meant NZ helped their international affairs. The people of Tokelau keep voting to stay part of NZ. The Federated States of Micronesia, Marshall Islands and Palau are in free association with the US.

The United Nations keeps a list of non-self-governing territories. The list includes Pacific countries. France still has the colonies of Wallis and Futuna, New Caledonia and French Polynesia. The US has Guam, Northern Mariana Islands and American Samoa. Hawaii was taken off the list when it became a US state.

Draw a cartoon strip to show the meaning of decolonisation.

List all the Pacific Islands mentioned in this unit. Find out about how they are governed today.

Political Unrest

 RESOURCE 1   Fiji

Fiji's ethnic groups are Fijian 51%, Indian 44%, other 5% (eg European, other Pacific Islanders, Chinese). The British brought Indians from India to work on sugar plantations. This caused tension between Fijians and Indians.

Some Fijians became unhappy when the constitution gave Indians equal rights with Fijians. In 2000 a gang led by George Speight stormed parliament. They kidnapped the Prime Minister and 35 other MPs. George made himself Prime Minister. He said he had got rid of the constitution. The gang holed up in the parliamentary complex. Some soldiers joined the rebels. But George Speight and many other rebels were arrested.

1   Draw a graph to show the ethnic groups in Fiji.

2   Practise saying the word 'coup'. It is a French word and you do not sound the 'p'. It is short for coup d'etat. It means a sudden, often violent, overthrow of a government such as George's coup d'etat.

3   Choose one word (4 in total) to describe how each of the following may have felt towards the coup: George Speight, the kidnapped Prime Minister, the soldiers who joined the rebels, the soldiers who arrested George.

4   A constitution is rules for how parliament is organised. Explain the probable link between the coup and the constitution.

 RESOURCE 2   Cartoon

1   Identify the 3 people.

2   Comment on the clothes.

3   Explain what is happening in the cartoon.

4   Invent a name for the figure on the right based on what that figure is saying eg Mrs Wise because she guessed the future.

5   Decide which country the cartoon is set in and give a reason for your decision.

 **RESOURCE 3**   **Bougainville**

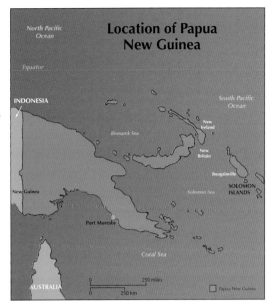

Location of Papua New Guinea

The culture of Bougainville is closer to Solomon Islanders than the rest of Papua New Guinea. Bougainville rebels fought to get free of PNG. Thousands died. In 2005 Bougainville won some self-government. It also got the promise of a future vote on whether it will become independent from PNG.

The rebels closed the giant Panguna copper mine. It was the world's richest. The PNG government, supported by Australia, responded to the action of the rebels by sending in riot police and military, to try to re-open the mine. They failed.

| 1 Write a sentence about the location of Bougainville in relation to the rest of PNG. | 2 Quote the sentence that offers a reason for the rebels wanting independence. |

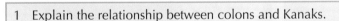 **RESOURCE 4**   **New Caledonia**

The French sent thousands of convicts to New Caledonia. Today the French there are called colons. The original inhabitants call themselves Kanaks. They call New Caledonia Kanaky. There are slightly more Kanaks than colons. The Kanaks want to be free of France. They say France stole their land. Violence erupted in the mid-1980s. France declared a state of emergency and sent paratroopers. France finally agreed to a vote on independence, to be held in 1998. But later France said the vote would be held between 2014 and 2019.

1 Explain the relationship between colons and Kanaks.

2 Account for why countries like France shipped convicts off to the Pacific.

RESOURCE 5   **Tonga**

Tonga is the only monarchy in the Pacific. The Royal Family has a lot of control over media and parliament. Tonga has a traditional society. It is very Christian. Some young, Western-educated Tongans want a new, more democratic constitution. Recently there have been strikes and street protests.

Tonga's Royal Palace.

1 Quote the sentence that shows action by people demanding political reform.

2 Explain how some Tongans want to change traditional society.

3 Draw symbols to show what the following are: monarchy, media, democratic.

RESOURCE 6 — Solomon Islands

There has been a lot of fighting between groups. In 2003 Australia, NZ and other Pacific governments answered the Solomon request for help. They sent troops to stop the country becoming lawless. In 2006 elections were held. One group said the vote was fixed. They said the new PM was in the pocket of the Taiwanese. They accused him of accepting millions of dollars in kickbacks in return for recognising Taiwan over China. They looted and torched about 90% of Chinatown. Most Chinese fled the country. Schools, shops, banks and offices closed. NZ and Australia sent extra troops to get order.

1 Quote the key sentence that shows the Solomons is not a united country.

2 Explain why troops from your country have been sent to the Solomons.

A hotel gutted by fire after protesters burned it down, 2006.

New Zealand troops were sent to the Solomon Islands to help control the unrest, 2006.

 USE THE NET

1 Find out what happened to George Speight after his arrest.

2 There has been tension between Australia and Norfolk Island. Find out what the present situation is.

3 Find out what the present situation is in the Solomons and in Tonga.

4 Find out how New Caledonia got its name.

UNIT 14 The Pacific Forum

> The word forum comes from Latin in the days of Ancient Rome. It means a meeting for the open discussion of topics of public interest.

RESOURCE Data

What

It is a group of countries working together like a Pacific Islands *extended family*.

Why

Pacific countries are small. They know they need to act together to be a strong voice. They will be able to use their resources more *sustainably* if they work together. They aim to

- keep the peace in the Pacific and make the Pacific free from crime
- help countries be good regional neighbours
- help countries be good international *citizens*
- improve the economic situation of Pacific countries
- make the Pacific free of big environmental problems and look after *natural resources*.

Who

Members are Australia, NZ, Cook Islands, Fiji, Nauru, Tonga, Samoa, Niue, Papua New Guinea, Kiribati, Tuvalu, Vanuatu, Solomon Islands, The Republic of the Marshall Islands, The Federated States of Micronesia, Palau. All member countries meet as equals.

When

It was set up in 1971.

Where

Forum *Heads of Government* meet in one member country for several days each year.

How

There are no rules about how things are done. Meetings are *informal*. Decisions are made by general agreement. This system is known as *The Pacific Way*.

NZ and Australia each pay about one third of the cost of the Forum's budget. Pacific Island countries *collectively* pay the other third.

Examples of achievements

- The *South Pacific Nuclear Free Zone* in 1985 (bans making, owning, setting up and testing of nuclear explosive devices in treaty territories; also bans dumping nuclear waste in the Pacific).
- The *Biketawa Declaration* (says Forum members will help each other in time of crisis or if a member asks for help).

1 Create 10 questions about this resource. The answers are the words in red.

2 Draw a graph to show who pays the Forum's bills.

USE THE NET

Find out about the two agreements mentioned.

Chequebook Diplomacy

Diplomacy refers to how countries relate to each other in a friendly way. Chequebook diplomacy refers to countries getting friends by paying big sums of money.

RESOURCE 1 — China and Taiwan

Taiwan is an island off the coast of east China. It has a non-communist Chinese government. Mainland China has a Chinese communist government. China and Taiwan both claim to be the one true China. Each wants other countries to recognise it as the rightful China. China says that Taiwan is simply a province of China. It plans to unite China and Taiwan. Taiwan says that will never happen.

Taiwan and China have been trying to make friends in the Pacific. For example, recently the Chinese Premier flew to Fiji for the first China-Pacific Island Countries Economic Development and Cooperation Forum. He was generous to the countries which did not recognise Taiwan. He gave out $615.54 million in loans. He removed import tariffs. He cancelled debts. He promised free malaria medicine. He added Papua New Guinea, Samoa and the Federated States of Micronesia to the list of places Chinese were allowed to visit. China had already got Fiji, the Cook Islands, Vanuatu and Tonga onside. Taiwan had got Kiribati, Nauru, the Marshall Islands, Palua, Solomon Islands and Tuvalu onside.

1 Make a labelled sketch to show where China and Taiwan are located in relation to each other.

2 Find evidence of possible chequebook diplomacy.

3 Suggest a reason why China welcomes Pacific leaders to China with a lot of ceremony.

4 Suggest a reason why much of Chinese and Taiwanese help in the Pacific is in the form of expensive official buildings.

RESOURCE 2 — Effects on the Islands

Support of China or Taiwan over the other can cause conflict in the Pacific countries. For example, recently the Vanuatu Prime Minister said he was switching diplomatic recognition from China to Taiwan. His government rebelled and kicked him out. Vanuatu went back to China.

The Pacific countries have to find friendly solutions. For example, China got membership to the Council of South Pacific Tourism Organisation. It is a regional body based in Suva. Taiwan wanted membership too.

Recently Fiji let Taiwan's President travel through Nadi on his tour of Pacific countries which recognised Taiwan. China was angry. Fiji said the Pacific way is to welcome people.

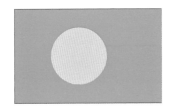

Taiwan has spent millions on Tuvalu such as the new 3-storey government building. When the Prime Minister of Tuvalu made a trip to China, he lost power in Tuvalu. People were angry at him for putting Tuvalu at risk.

The number of Chinese going to live in Fiji has made some in Fiji worry about links to drug trafficking, illegal gambling, and violent crimes.

1 List good and bad things about Pacific Islands being 'chased' by China or Taiwan.	2 With your group try problem-solving the China/Taiwan situation in the Pacific.

RESOURCE 3   Views

China says it wants to help because China has received help from developed countries in the past. It says it is the biggest developing country and has an obligation to offer help to others.

Taiwan wants a formal relationship with every country in the world. It wants other countries to recognise it as a separate country from China. It wants countries to support its efforts to join the United Nations.

The European Union has 25 member countries. The EU is interested in the Pacific region. It wants to develop sustainable management of natural resources such as timber. China could be a threat to natural resources in the Pacific region. If China helps deforestation in the Pacific, this may one day help climate change in the northern hemisphere. If Chinese money goes to dishonest officials or the buying of weapons, it could increase conflict there.

Japan is also putting a lot of money into the Pacific. At a recent international whaling meet, the Solomon Islands, Kiribati, and Tuvalu voted with Japan to help get commercial whaling legal again. This was against countries such as NZ and Australia who voted against it.

Anti-whaling protest.

1 Explain the difference between a developed country and a developing country. 2 Work out the link between the EU and China in the Pacific. 3 Quote the sentence about using resources wisely so they do not get all used up.	4 Draw 4 speech bubbles. Inside them put a comment from a Chinese, a Taiwanese, an EU member, and a Japanese about the relationship between their country and the Pacific region.

USE THE NET

1 Find examples of help given to Pacific Island countries by China and Taiwan.
2 Find out to which country each flag shown in the unit belongs.

UNIT 16 Aid

RESOURCE 1   Bilateral and Multilateral Aid

aid = help given eg a rich country giving money to a poorer country	bilateral aid = aid given by one country to another country	multilateral aid = aid given by many countries, such as a group like the United Nations, to a country

NZ and Australia have leadership roles in the Pacific. They are neighbours of the Pacific Islands. They trade with them. They want to help keep the region politically stable. Many Pacific Islanders live in NZ or Australia.

In 2006 Australia gave $1118 million in aid to the Pacific region. Japan, China, Taiwan, the US, the EU, the Asia Development Bank, the IMF, the World Bank, and UN agencies also gave aid.

1   Name 5 bilateral aid givers.   2   Name 5 multilateral aid givers.

RESOURCE 2   The Target

The OECD stands for Organisation for Economic Cooperation and Development. Members are industrialised developed countries. NZ and Australia are members.

The OECD says developed countries should give 0.7 per cent of their gross national income to overseas aid. 0.7 per cent is 'the target'. Developed countries do not reach this target.

In 2005 the NZ Government increased its overseas aid budget by 21 per cent. This was a boost of nearly $60 million. More than half the extra money went to the Pacific. The increase took the NZ foreign aid budget to $345 million. This was 0.27 per cent of NZ's gross national income. Bob Geldorf described NZ's aid as pathetic.

1   Explain the words in red.

2   Not many people can remember what OECD stands for. Invent a good way to remember it.

3   Account for the fact that none of NZ and Australia's Pacific neighbours are members of the OECD.

4   Create a graphic or cartoon of the OECD target and how develped countries do not reach it.

5   Find out who Bob Geldof is.

6   Explain why the pictures could have been chosen for this unit.

Tokelau — rugby field. Tokelau is a territory of New Zealand.

RESOURCE 3   Economic Growth

A recent report by the United Nations said economic growth in the Pacific was averaging less than 5 per cent. One cause was corruption, law and order problems, and bad governing. These things made an unfriendly environment for investors. It said the Pacific must get sustainable development and stop natural resources being wasted. It said tourism was good for economic growth.

Experts say islands should do other things besides accepting aid. They should look for foreign investment. They should change communal land ownership. This would encourage people to improve their own piece of land. They should reduce birth rates. They should bring in some privatisation. They should downsize their governments. Guam holds the over-governed world record.

1 Find the word that means taking control of a resource out of government hands and putting it in a private company's hands. Learn how to spell the word.

2 Explain exactly what Guam holds the world record for.

3 List advice given to Pacific islands to help improve economic growth.

USE THE NET

Find the latest figures to show how much your country is giving to the Pacific in aid. Try to find details about where the aid is going.

Tokelau — school house.

Tokelau — loading a passenger barge.

Conservation in Marine Reserves

RESOURCE 1   Conservation

Conservation is about looking after the environment and its resources. It aims to make sure you get to use and enjoy the environment and its resources. In return you must look after them and never use them all up. The term for this is sustainable use. It is about using a resource in a way that does not destroy it. This means that your children will be able to use and enjoy the same resources.

1   Decide if you can use a resource while you are conserving it. Give evidence to back up your answer.

2   'Sustain' means to keep up or keep going. Explain what sustainable means.

3   Learn how to spell these terms: environment, conservation, resource, sustainable. Work out which word most often trips people up when they try to spell it.

 ## RESOURCE 2   Newspaper Report

The tiny Pacific Island nation of Kiribati has announced the creation of the world's third-largest marine-protected area.

The statement was made at a UN-sponsored <u>environmental</u> conference in Brazil.

The <u>protected</u> area at the Phoenix Islands, located about half way between Fiji and Hawaii, places 191,141.26 sq km off limits for commerical fishing, protecting <u>precious</u> coral reefs and undersea mountains.

"If the coral and reefs are protected, then the fish will thrive and grow and bring us benefit," said Anote Tong, President of the republic of Kiribati.

Tong also said his nation, the Federated States of Micronesia and the republic of the Marshall Islands would work to protect 30 per cent of the nearshore <u>marine</u> areas and 20 per cent of land resources on islands by 2030.   [AP 30/3/2006]

1   Give the year in which the marine reserve was announced.

2   In not more than 15 words explain where the marine reserve is.

3   Account for why commercial fishermen would be interested in this news.

4   Decide if the report says exactly how the area is to be protected. Give a piece of evidence to back up your answer.

5   Name the marine features Kiribati wants to conserve.

6   Give a direct quote in the report.

7   Name the 2 other Pacific countries with which Kiribati plans to work to protect other areas by 2030.

8   Explain why the writer chose to use the word "tiny" (in the opening sentence).

RESOURCE 3 Kermadec Islands Reserve

The Kermadec Islands lie 1000 kilometres northeast of New Zealand. They are a chain of islands and rocks. They lie on the edge of the Kermadec Trench. This is where the Pacific Plate buries under the Australian Plate. The islands are the tops of volcanoes. These volcanoes rise 8000 metres from the sea floor. There are volcanic eruptions, earthquakes, and the occasional cyclone. Raoul is the largest island. In 2006 a crater lake erupted there. It killed a Department of Conservation worker from NZ.

The Kermadecs have a marine reserve of 745,000 hectares or 7450 square kilometres. It goes out 12 nautical miles from the shallows inhabited by the rare spotted black groper to the deeper areas of the Kermadec Trench.

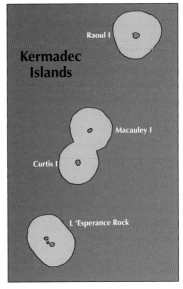

Kermadec Islands

Raoul I
Macauley I
Curtis I
L 'Esperance Rock

1   The map shows the main islands of the Kermadecs. Give their names.

2   Copy the map and add a title, a key (legend), orientation (north direction).

3   Suggest a reason the Kermadecs are not inhabited, except for workers who visit sometimes. Give a piece of evidence to support your reason.

4   Give 2 facts about the spotted black groper.

RESOURCE 4 Marine Reserves in the Cook Islands

There are 5 marine reserves/ra'ui in the lagoon on Rarotonga. They are Tikioko, Aroko/ Nukupure, Matavera/Pouara, Nikao, Kavera. They aim to protect the marine environment. Plants and animals get the chance to grow and spread to other parts of the lagoon and sea. The total reserve area is about 8% of the Rarotonga reef. You can swim and snorkel in those areas. Taking marine life, especially traditional food, is restricted or banned.

1   Give the term used in the Islands for a marine reserve.

2   Write a sentence about the link between the reserves and conservation.

3   Explain the difference between 'restricted' and 'banned'.

4   Decide if glass-bottom boat cruises are likely to be popular in the reserves. Give a piece of evidence to back up your answer.

USE THE NET

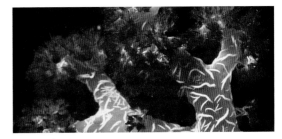

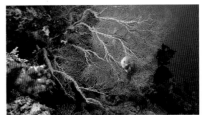

1   The 2 marine reserves larger than the Kiribati one are Great Barrier and Northwestern Hawaiian Islands. Find out about them and do a short profile on each one. Look especially for points of similarity and difference with Kiribati.

2   Find out 10 facts about the Kermadec Islands.

3   Choose another 5 Pacific countries. You can include your own country. Find out attitudes to conservation in each of your 5 countries.

UNIT 18 Fisheries

RESOURCE 1   Biodiversity

The Pacific Ocean's biodiversity (its range of species) is huge. The most commonly shared fish stock is tuna. Also important are species like orange roughy, alfonsino, oreo doreis and jack mackerel. Mahi-mahi means strong in Hawaiian. It is also called el Dorado and dolphinfish. It is fast growing. *This makes it good for sustainable fishery.*

The world's largest fish, the whale shark, lives in the Pacific. So does the world's largest turtle, the Pacific Leatherback. As does the world's fastest fish, the sail fish. It can swim at 110 km an hour.

There are dangerous fish. Rabbitfish and Scorpion fish are mainly in reefs. Their spines can give painful cuts. They could even kill you. The blue-ring octopus is usually on Great Barrier Reef off eastern Australia. Its bite can kill.

Pogonophora worms inhabit the seabed near vents (cracks) where water comes into contact with molten earth.

The sea has always been a major source of food for Pacific Islanders. Before food was imported into the Pacific, protein came from seafood. Even today no seafood is wasted. Jellyfish, sea slugs, sea cucumbers, seaweeds, octopus and the tiny coral worm palolo are among seafood delicacies Pacific Islanders eat. Seafood such as sea cucumbers have also been used as medicine.

1 Rewrite the sentence in italics to explain exactly what it means.

2 Find a piece of evidence you would show somebody who said no life could exist in the depths of the Pacific.

3 Describe or draw the location of the Great Barrier Reef.

4 Create a chart to show the biodiversity of the Pacific Ocean.

RESOURCE 2   Pearl Fishing Farms

Black pearls are about 100 times rarer than white pearls. The oysters that produce them grow up to 30cm in diameter. So the pearls are generally much larger than white ones. Necklaces of a dozen or more top-quality black pearls cost 6-figure sums.Legend says that the Polynesian god of war and peace, Oro, came down to Earth on a rainbow to offer this special pearl oyster to mankind.

French Polynesia and the Cook Islands are the best known places for black pearls. Other Pacific Islands such as Hawaii and Fiji are trying to break into the market.

1 Identify the 'trick' letter in each of the following words: pearls, special, islands.

2 Give the reason suggested for black pearls being so expensive.

3 Pearl farms come under the category of aquaculture – the cultivation of marine resources for humans to use or eat. Work out a) how the word aquaculture shows what it means, and b) whether or not pearl farms would be a sustainable use of the ocean.

Traditionally people took only as much from the sea as they needed for food. This is called subsistence farming. Today's technology allows huge catches. Technology lets the catches be carried away to other countries. Technology brings tourists from other countries. They like to eat traditional fish dishes.

Overfishing of any species upsets the balance among species. The trochus shell is the main food of the crown of thorns starfish. The triton shell is the enemy of this starfish. Both shells have been overfished. So the crown of thorns has turned to coral for food.

Crown of thorns starfish.

1   Traditional Pacific societies looked after marine resources. If an area showed signs of overfishing, it was declared tapu until stocks rebuilt. Suggest why such traditional conservation methods are not always used today.

2   Explain the links among the 3 species mentioned in the last paragraph.

3   Give details of the connection between technology and tourists.

4   Make up a graphic to show the difference between subsistence fishing and non-subsistence fishing. Use as little writing as possible on your graphic.

Traditional fishing.

USE THE NET

1   Find out how important fishing is to your country's economy.

2   Scientists are carrying out a world marine census. On average they are discovering 3 new fish species a week. By the time they finish in 2010, they may have found more than 2 million different species of marine life. Find out about the census, especially in the Pacific.

3   Aquaculture can take pressure off natural marine resources. Some islands are experimenting with farming fresh or brackish water species. Find 2 examples.

4   Many Pacific nations have agreements with larger nations to develop their marine resources. Find 2 examples.

A Big Issue - Whaling

RESOURCE 1   Fishing and Whaling

Most fish breed quickly. They reproduce by releasing huge numbers of eggs into the water. Whales are mammals. They have only a single calf at a time. The calf needs at least a year of mother care before it can survive on its own. It will not reproduce for many years.

> The resource is the answer to a question that many people ask about fishing and whaling. See if you can work out what the question is.

RESOURCE 2   Newspaper Report

TOKYO – Whale meat may soon be on the school menu and served at family restaurant chains as a new Japanese company tries to win over younger consumers.

The move is likely to outrage anti-whaling nations and environmental groups, who have long charged that Japan's programme of what it calls research whaling is really commercial whaling in disguise.

Tokyo, which maintains that eating whale is a treasured cultural tradition, abandoned commercial whaling in 1986 in line with an international ban. But it began research whaling the next year for a return to limited commercial whaling.

[12/5/2006]

1 What is a consumer?

2 What reason does Japan give for offering whale meat to school children?

3 Why would there have been an international ban on whaling?

4 In what year did Japan begin research whaling?

5 What does 'outrage' mean?

6 What does 'abandoned' mean?

7 Why did Japan stop commercial whaling in 1986?

8 What would Japan say is the difference between commercial and research whaling?

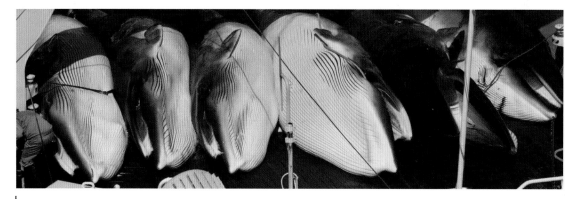

RESOURCE 3 Japan and the Pacific Nations

Australia and NZ, Solomon Islands, Kiribati, Tuvalu, Palau and Nauru belong to the IWC (International Whaling Commission). They vote on whether or not to expand scientific and commercial whaling. Japan catches hundreds of whales every year, mainly Minke whales from the South Pacific.

Some non-IWC nations in the Pacific are against Japan's whaling. Tonga encourages South Pacific nations to create ecotourism ventures around whale watching. Whales in Tonga are protected by a Royal decree from the King. Tahiti says French Polynesian waters are a haven for many species of marine mammals which play an important role in whale reproduction. The Cook Islands, French Polynesia, Niue, Australia, and Fiji have declared whale sanctuaries. NZ and Vanuatu have laws that protect whales within territorial waters.

1 Why does the IWC hold votes at its meetings?

2 What does 'decree' mean?

3 What are the 2 words in Resource 3 that mean a refuge and keeping safe?

4 What is an ecotourism venture?

5 Would whale watching tours fall into the category of scientific whaling, or ecotourism ventures? Give a reason for your answer.

RESOURCE 4 Pictures

Past

Present

A B

1 Describe differences you can see.

2 Decide which group of whalers, A or B, would have understood most about biodiversity and sustainability. Give reasons for your answer.

3 Whaling is an emotional issue today. Explain what emotional means and why whaling is an emotional issue.

USE THE NET

1 Many Pacific peoples traditionally had a special relationship with whales. For example, a legend in the NZ film *Whale Rider* has a whale bearing the tribal ancestor, Paikea, to NZ after his canoe sinks on the voyage from Hawaiki, the ancestral home of Maori. Find out another Pacific legend about whales.

2 Find out why whales have been hunted.

3 Australia and New Zealand have proposed a South Pacific Whale Sanctuary. Find out what progress has been made on this.

UNIT 20 Examples of Environmental Issues

RESOURCE 1 Ecosystems

| **ecology** = study of plants and animals interracting with their environment | **eco** = to do with ecology | **system** = a group of things forming a whole |

The ecosystems of the Pacific are some of the most amazing on Earth. Firstly there is the ocean environment — 29 million square kilometres of it. Then there are the land ecosystems strung out across the ocean. They can be difficult to survive in. They are often fragile. Coral, for example, might give you a nasty nip if you stand on it but you might damage it in return. Yet people have lived in these ecosystems for millennia. They have survived disasters that nature hurled at them. But now they face man-made disasters.

Use the material in the box to help you describe or draw what an ecosystem is.

RESOURCE 2 Endangered Species

Endangered refers to plants and animals that are in danger of becoming extinct. They need protection to survive. When outsiders came into the Pacific they brought diseases and different animals and plants. These affected the biodiversity of the region. For example, 9 of the 12 native forest bird species of Guam are extinct. Biologists said the introduced brown tree snake is mainly responsible. Endangered marine species in the Pacific include dugong, sea lion, sea otter, seal, leatherback turtle, and grey whale.

Dugong.

1 Create a graphic to show the difference between extinct and endangered.

2 Give the 2 reasons stated for species being endangered. Suggest other reasons.

RESOURCE 3 Deforestation

Deforestation is the cutting down of trees in forests. In the past, Pacific people cleared trees for subsistence agriculture, and to get wood for fuel and building. Today deforestation is increasing. Big companies from places such as Japan and South Korea have found timber resources in PNG, Solomon Islands and Vanuatu. Logging in the Solomons, for example, is already at unsustainable levels. Yet it is the rich companies, not the people of the Solomons, who are getting richer.

1 Explain what happens when things continue at unsustainable levels.

2 Give details about why it matters that logging is at this level. Think of plants and animals.

RESOURCE 4 Nauru

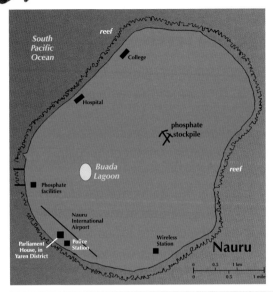

Nauru is a tiny phosphate rock island. The raised plateau is in its centre. The highest point is 61m above sea level along the plateau rim.

Phosphate built up from thousands of years of sea bird droppings. It has been mined by a German-British group, and a British, Australia and NZ group. In 2005 an Australian company made an agreement to mine the remaining supplies.

Mining has left the central 90% of Nauru a wasteland. Nobody can live or grow anything there. Living on processed food has led to obesity and diabetes.

1 Describe the shape of Nauru.

2 Work out what the 2 small horizontal lines on the left side of the island's reefs are most likely to be. Give a reason for your decision.

3 Work out what the long line beside Nauru International Airport stands for.

4 Make up a key / legend for the map.

5 Draw a plateau.

6 The people of Nauru have made money from phosphate. Describe the price they have paid.

RESOURCE 5 Dumping of Nuclear Waste

The world is running out of places to store nuclear waste. Russia, for example, recently accepted the 1993 ban on the dumping of radioactive wastes in the Pacific Ocean. Australia's wastes are now stored at over a hundred sites around Australia. In 2004 the Australian Prime Minister proposed sending Australia's nuclear waste to an offshore island. Storing nuclear waste on a Pacific atoll or in a country like Nauru is dangerously ridiculous, said the shadow industry minister. Export of Australian waste to Pacific Forum countries is banned by a treaty.

1 Find in the resource a difference of opinion about disposal of nuclear waste.

2 Explain the job of a shadow minister.

3 In your group talk about the problem of nuclear waste – where it comes from, where it should go, whether Pacific Island countries should agree to be dump sites or let barges of nuclear waste travel across the Pacific. Present findings to the class.

1 Wind energy might be an important source of energy to Pacific Island countries. Find out what the present situation is.

2 Choose an island and find out its special environmental problems eg Tonga has deforestation, damage to coral reefs from starfish and people collecting coral and shell, and the native sea turtle threatened by overhunting.

Rising Sea Levels

 RESOURCE 1 **Monitoring Rising Sea Levels**

In 1987 the United Nations set up an Intergovernmental Panel on Climate Change (IPCC). It was a group of 700 experts using computer models. They say Earth's temperature and sea levels are rising.

The South Pacific Sea Level and Climate Monitoring Project was developed as an Australian response to concerns raised by Pacific Island countries about climate and sea levels in the Pacific. It wants to help countries understand sea level rise.

The South Pacific Regional Environment Programme and the National Tidal Facility at Flinders University in Australia set up sea level monitoring stations across the Pacific. This made the Pacific region the first in the world to measure changes in sea level with good accuracy. Their results show a sea level rise of up to 25 millimetres a year. This is well above the IPCC global estimate. Satellite data backed up these findings. It has shown a 20-30 millimetre a year sea level rise in a region stretching from Papua New Guinea southeast to Fiji. That may not seem much of a rise. But it can cause a lot of flooding and sinking.

1 State 3 facts about the IPCC.

2 Suggest a reason why these organisations have such long names.

3 Draw a diagram to show what is meant by rising sea levels.

Fakaofu Island, Tokelau, is a low-lying coral atoll in danger of rising sea levels.

 RESOURCE 2 **El Nino**

Some scientists say a major cause of rising sea levels is El Nino. This is an abnormal weather pattern that happens every few years. El Nino is the Spanish name for little boy. Usually it happens at Christmas. So, little boy is to do with the Christ child.

El Nino is a change in temperature of sea currents. The change goes from cold to unusually warm. It happens off the west coast of South America. A huge area of the Pacific becomes warmer because of El Nino. Warm water rises above cold. So the sea level rises. The change in temperature and sea level changes wind patterns. The warm water travels across the Pacific from west to east. It takes rain with it. Scientists do not really understand how El Nino forms. But they say the storms it brings are getting worse.

1 Learn the spellings of scientists, El Nino, weather, pattern, abnormal, Christmas, currents, temperature.

2 Discuss with the class why it is important you learn about El Nino.

3 Work in a group to produce a graphic about El Nino. You could do some extra research. As a class, decide which group's graphic is the best.

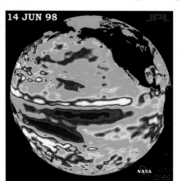 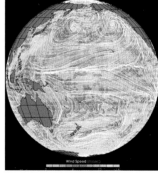

Satellite image of Pacific during El Nino. White and red areas are up to 30 cms higher than normal.

RESOURCE 3   Global Warming

global = to do with the globe
globe = Earth, the whole world
emission = discharge, something given out
greenhouse effect = process of warming
    which makes life on Earth possible
carbon dioxide = an important greenhouse
    gas
enhanced = made bigger, stronger
enhanced greenhouse effect = when humans
    create more carbon dioxide through
    activities such as putting factory emissions
    into the atmosphere, the greenhouse
    effect becomes stronger

Scientists talk about the enhanced greenhouse effect. They say it is making the world warmer. A warmer world will have a higher sea level. This is because when land and lower atmosphere warm, heat is transferred into the oceans. When materials are heated they expand. So the transferred heat causes sea water to expand. This results in a rise in sea level. Water from land ice such as melting glaciers may enter the ocean. This adds to the rise.

Sort out the following into causes and results of global warming: spreading of tropical diseases, losing food-producing areas, burning rubbish outside, increasing bad weather, volcanic eruptions, rising sea levels, animal emissions, open fires, drowning coastal areas, vehicle emissions, changing habitats for plants, clearing forest areas, using chlorofluorocarbons in electronic appliances, industrial emissions

RESOURCE 4   Cartoon

1   Work out how long ago it was published and compare the date of publication with your date of birth.

2   Present evidence to show from which country the fishing boat comes.

3   Suggest a reason you know that the men are not scientists.

4   Explain what the man on the left is doing.

5   Supply a more scientific term for 'hotting up'.

6   Comment about whether the cartoon is about an issue that is still important today.

In 2001 Australia had the highest level of greenhouse gas emissions per person of any industrialised country. Find out the present list of top 10 contributors to greenhouse gas emissions.

UNIT 22

Effects of Climate Change and Rising Sea Levels

RESOURCE 1   Low Islands and High Islands

Nations such as Kiribati, Marshall Islands, Tokelau, and Tuvalu have land that is less than about 4 metres above mean sea level. Rising sea levels could sink their islands.

Countries with high islands might think a rising sea level is not their problem. But most of their people live along coasts. Many coastal plains already have erosion (land being eaten away) and flooding.

Most commercial activity is located along coasts. For example, Fiji's sugarcane is grown mostly on coastal flat areas. Sea water has been getting into coastal aquifers (layers of rock which hold water). This lowers sugar content in the cane.

Tourism is an important source of money for Pacific Islands. Tourists want sun, sand, sea. Until now, beaches have looked after themselves. But sea level rises are eroding tourist beaches. Some resort owners now have to spend a lot of money looking after their beaches.

1   Suggest 2 reasons for crops being grown along the coast.

2   Identify the reason for research being done to develop strains of sugarcane, and other crops, which cope better with salt.

3   Compose a sentence with the word 'erosion' in it and another sentence with the word 'eroding' in it.

4   Aquifers are precious resources. Decide if you agree with this statement and be prepared to explain your reasons.

5   'Intrusion of salt water' is a term scientists and geographers use. Work out where the word 'intrusion' comes from.

RESOURCE 2   Food Crops

For millennia, gardens have been a key to the survival of Pacific Islanders. Where soil is poor, or where there is little or no soil, people plant crops in pits of compost. Once the sea gets into them, pits and gardens are ruined. Rising sea level is changing traditional gardening. Many people in Funafuti in Tuvalu, for example, now grow taro in old kerosene cans.

List the advantages and / or disadvantages (good and bad things) of using kerosene cans for crops.

RESOURCE 3   Water Supplies

Climate change is bringing droughts. In one recent year Fiji's sugarcane production fell by two thirds. Tonga's squash crop was more than halved. In Papua New Guinea, Australia spent millions of dollars delivering food to isolated areas of highlands and low-lying islands. Many people there were close to starvation. In Micronesia almost 40 atolls ran out of water; the capital, Pohnpei, had to use brackish underground supplies.

Rising sea levels will invade freshwater aquifers. This will ruin drinking water.

1 Use 2 stick figures and speech bubbles to show the difference between fresh water and brackish water.	3 Explain the link between drought and loss of production.
2 Invent a symbol you could use to stand for 'drought'.	4 Create a star diagram to summarise the information in this resource.

 RESOURCE 4   **Fisheries and Ocean Currents**

Traditional knowledge of the currents and the best fishing places are handed down through generations. Global warming is changing the location of these.

1 A word used by scientists and geographers is 'adapt'. It means having to change your thinking and habits because your environment is changing. Explain how global warming may force people to adapt.	2 In your group plan a dialogue (speaking) between the boy and the fish in the picture. Present it to the class.

 RESOURCE 5   **Diseases**

Global warming will affect the health of Pacific Islanders. Places that were too cold for mosquitoes which carry disease will be warm enough. Malaria and dengue fever have already started to spread. Diseases caused by food poisoning and contamination of drinking and swimming water could rise. Sick people can lead to a breakdown in sewerage and rubbish collections which can lead to other diseases such as cholera. Heat waves can cause illness. An increase in natural disasters such as floods could increase stress and feelings of sadness.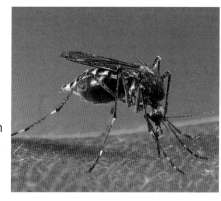

1 Explain how global warming could affect people's physical health.	2 Explain how global warming could affect people's mental health.

 RESOURCE 6   **Coral Reefs**

Coral reefs surround many Pacific island groups. They are a habitat for sea creatures the islanders catch to eat. The reefs also protect the islands from huge waves stirred up by storms or undersea earthquakes.

Rising sea levels and warmer waters put stress on reefs. Warmer waters can not sustain large fish populations. Reefs may suffer bleaching. This happens when reef-building corals loosen their algae. The algae help feed the reefs. They give colour to the reefs. The starved corals look pale. Continued bleaching eventually kills coral. Many islands are reporting bleaching of their coral.

Bleached coral.

1 Draw a sea creature. It could be a fish or something you invent. Now draw and label its habitat. Put a speech bubble from your creature's mouth. In the bubble, put its comment about its habitat.	2 Compare the habitat you live in with the habitat that a coral reef provides. 3 Explain where the term 'bleaching' came from.

"We are waiting to see concrete evidence of global warming."

"Rich developed countries who have contributed to global warming should fix the problem."

"We have no industry. We have a small population. We don't contribute much to global warming."

"Rich developed countries who have contributed to global warming should pay for the effects on Pacific Islands."

"If rich developed countries do not stop global warming, they will be responsible for killing Pacific people."

"Some scientists say there is no rise in sea level. But the tide is rising. We see it with our own eyes."

"Reports of global warming and rising sea levels are over-exaggerated. They pose no threat to people."

"Pacific Islands are still recovering from the nuclear bomb when rich developed countries carried out tests in our environment. Now we face the climatic bomb."

"Rich developed countries are denying future generations of Pacific Islanders their right to live where their ancestors have lived for thousands of years."

"Rich developed countries don't care if Pacific Islands sink or fly."

1   Controversial means that not everybody shares the same attitudes about an issue. Use evidence from the resource to show that this issue is controversial.

2   Sort out some things about the issue that people have differing attitudes to.

3   In groups suggest ways this issue could be made less controversial.

USE THE NET

1   Make a list of tropical diseases.
2   Choose one disease and prepare a fact sheet about it.
3   Find evidence to show that coral reefs are amazing habitats.

Examples of Islands Hit by Climate Change

RESOURCE 1   Marshall Islands

Majuro (the capital) has lost a lot of its beachfront already. Locals believe storms and high tides are causing this. Majuro has built sea walls to stop further erosion. It has even used garbage imported from the US. But the walls need constant rebuilding. This will possibly cost more than the Marshall Islands' annual budget. The highest point on the islands is about 10m. The US used some of the atolls to test their nuclear bombs on. Some islanders were forced to leave their homes. The Marshalese worry that history will repeat itself.

1   Give the name for a person from the Marshall Islands.

2   Decide what the last sentence means. Give a piece of evidence from the resource to back up your decision.

Marshall Islands sea wall.

RESOURCE 2   Kiribati

Satellite image of Kiribati coral atolls.

Kiribati is 33 spread-out coral atolls. People report unusually high tides and unusually strong storms with rogue waves. Sea water is intruding into fresh water. This spoils drinking water and destroys crops. Kiribati's oral history goes back thousands of years. It says that Tebua Tarawa, an islet, used to be a resting place for fishermen. They beached their boats and got coconuts to drink. Recently the coconut trees disappeared. Then the sand banks disappeared. Then the whole islet disappeared under the waves. On the islet of Bikeman, people used to present their gifts to the gods. Today people walk on the islet in knee-deep water.

1   Draw a graphic to show what a rogue wave is in comparison to a normal wave.

2   Find evidence in the resource to show rising sea level may affect culture.

RESOURCE 3   **Carteret Atoll**

Carteret Atoll is also known as the Kilinailau Islands or the Tulun Islands. It is a set
of 6 small coral islands. It is part of Papua New Guinea. Total land area is 0.6 square
kilometres. The highest point above sea level is 1.5 metres. 1000 – 1500 people live there.
They report recent increases in bad weather and rising sea levels. One storm washed away
a lot of shoreline. It cut an island in half. Breadfruit crops struggled. People depended on
aid drops from the government. A new sea wall was washed away. Seawater swept through
the islands. It left families homeless. It destroyed food gardens. In November 2005 the
Government decided to evacuate the islands. 10 familes are to go at a time. Experts say
that by 2015 Carteret Atoll will have sunk or will be largely underwater.

> Be a TV (or some new invention) reporter watching the last of the people of Carteret Atoll being
> evacuated. Make your report.

 RESOURCE 4   **Takuu**

Takuu, also in Papua New Guinea, is sinking. 2,500 people live there. They have little to
do with the outside world. They have kept their culture. They are known as The Singing
Polynesians because they can sing hundreds of songs from memory. Rising sea levels have
destroyed gardens. People face starvation. The older people who want to stay say they will
die when their island is underwater.

> 1   Explain the link between the third and   who went to Takuu and why your research
>   fourth sentences.   is important.
>
> 2   Give details of how you would go about   3   Suggest a reason for the older people's
>   studying the songs if you were a researcher   attitude.

 **RESOURCE 5**   **Tegua**

Tegua is an island in Vanuatu. By 2005 the settlement there, called Lateu, was being
flooded 4 or 5 times a year by high tides. The chief remembered as a young boy walking
30 metres from his house to fish from a rocky beach platform. Now the platform was
underwater. Coconut palms stood in water. So the people started dismantling their
wooden homes. They moved inland. The chief had to move first. They had to plant new
coconut palms and new gardens. The United Nations said Lateu has become one of, if
not the first, groups to be moved out of harm's way as a result of climate change.

> You are a villager on Tegua Island. Prepare a speech for a visitor from the United Nations saying
> how you feel about climate change.

USE THE NET

1   Prepare a fact sheet on Kiribati. Mention environmental problems.
2   Find out what the freshwater lens under the coral of islands is and how it
    works.
3   Find out why the people of Takuu have appealed to the world and what the
    situation is like there now.
4   Find out how the evacuation of Carteret Atoll is going.

UNIT 24 Environmental Refugees

refuge = safety, shelter
refugee = person who flees to a refuge ie runs away from own place which has become unsafe to another place which is safer

RESOURCE 1   What the United Nations Says

The number of environmental refugees is rising. The latest reports show that more people around the world are becoming refugees because of environmental disaster than because of war.

The best guess is that by 2050, there could be more than 150 million environmental refugees. Most will be due to effects of global warming. By 2006 environmental refugees were still not recognised as refugees under international agreements. So they could not legally get help.

Many people do not know what an environmental refugee is. Design a poster about environmental refugees.

Low tide, Funafuti, Tuvalu.

RESOURCE 2   Tuvalu

Tuvalu is one of the smallest and most remote countries on Earth. It used to be called the Ellice Islands. 'Tuvalu' means 'group of eight', referring to the country's eight traditionally inhabited islands. Pressure of population has led to people now living on the ninth island.

Tuvalu is a chain of 9 islands. 6 are coral atolls. 3 are limestone reefs. The highest point is 5m. There are no big towns, only villages. About 11,800 people live there.

The soil is poor. There are no known mineral resources. Subsistence farming and fishing are the main economic activities. Tuvalu gets a lot of money each year from an international trust fund set up by Australia, NZ, and the UK and supported also by Japan and South Korea. The US also gives money for a treaty on fisheries. In 2000, Tuvalu leased its internet domain name '.tv' for $50 million in royalties over the next dozen years.

Tuvalu has recently had increased environmental problems, such as water shortage. Australia, NZ, Britain and Japan have helped by providing desalination plants.

Tuvalu has always had poor soil. But the resource gives details about other features of Tuvalu and its society that are new. List these.

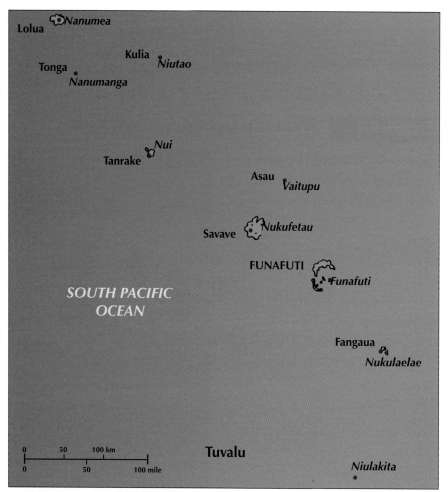

Lolua 　Nanumea
Kulia 　Niutao
Tonga
Nanumanga
Nui
Tanrake
Asau
Vaitupu
Savave 　Nukufetau
FUNAFUTI
Funafuti
SOUTH PACIFIC OCEAN
Fangaua
Nukulaelae
Tuvalu
Niulakita

0　50　100 km
0　50　100 mile

1　Work out what the names of the 9 islands are. Explain how you decided.

2　Suggest what the other names on the map are. Work out how you could find out if you were right.

3　Name the capital of Tuvalu. Explain how you worked it out.

4　Work out the name of the island that was not inhabited at the time the name Tuvalu was created. Explain your reasoning.

5　Give the reason Tuvalu is called a chain of islands.

RESOURCE 4

Tuvalu is concerned about global increases in greenhouse gas emissions. Their low islands put them at risk of rising sea levels. Best estimates say that a sea level rise of 20-40 centimetres in the next 100 years could make Tuvalu uninhabitable.

In 2000, the government appealed to Australia and New Zealand to take in Tuvaluans if their islands became uninhabitable.

NZ has agreed to accept the whole population. The scheme will possibly operate until 2050. This is about 75 Tuvaluans being relocated as refugees to NZ every year.

At the time of the agreement with NZ, the Executive Director of the Australia Institute, Dr Clive Hamilton, was quoted as saying 'NZ's agreement to accept refugees from Tuvalu driven from their homes by rising sea levels highlights the moral bankruptcy of Australia's refusal to ratify the Kyoto Protocol.' In your group, decide what he meant and whether or not you agree with it.

RESOURCE 5   One Side of the Story

People in Tuvalu say rising sea levels are driving them out. They say the cause of the rise is climate change. They point to the recent erratic weather - floods, droughts, and high level of tropical cyclones. They say salt water intrusion has affected traditional food crops and caused coastal erosion. There are no streams or rivers and freshwater sources are brackish. They use Tepuka Savilivili as an example. It is an islet. The sea has swallowed its huge sand banks. Its coconut trees have disappeared. The sea is moving over its remaining rock. People used to come out to Tepuka Savilivili if they had problems. They could spend time there to clear their heads. Soon there will be no islet to visit.

High tide, Funafuti, Tuvalu.

1 Tell a story about how you, as a Tuvaluan, once visited Tepuka Savilivili.

2 Explain why Tuvaluans may soon be called environmental refugees.

RESOURCE 6   The Other Side of the Story

Some experts say maybe there are other things making Tuvalu's environment worse. For example, around 6,000 people live in a small area on Fongafale. This area is not much bigger than the average city park in NZ or Australia. Building has increased recently. The Government has paved roads. Taiwan has built a new 3-storey administration building. On the other side of Fongafale, about 10 kms across the lagoon, is an uninhabited islet. It shows no sign of sinking.

Some scientists say the beach erosion is not caused by rising sea levels but the taking of sand for building. During WW2 the Americans cut down all the coconut trees and built a runway on Fongafale. They dug out large pits at the ends of the runway to get sand and coral. These pits went below sea level and are still there. Locals have turned them into pigpens and rubbish dumps. Various plans over the years to tidy them up have come to nothing.

King tides on Tuvalu and the sea washes inland.

1 Explain how this viewpoint differs from that of the Tuvaluans.

2 In your group, talk about the difference of viewpoints. Decide if, or how much, it matters and why.

1 Find out how the United Nations is involved in the world's environments.

2 Check if the UN has decided on an official definition of environmental refugees.

3 Find out what the present situation of Tuvaluans and NZ is.

Ideas in Pacific Society

RESOURCE 1 — Religion

Traditional societies had a strict system. Commoners were at the bottom. Chiefs were at the top. But Pacific culture in the past and today is full of differences. A custom on one island may not be the same on the next island. Even if the next island is part of the same country. What unites Pacific cultures today is Christianity. For example, Samoa's motto is 'Samoa is founded on God'. Churches throughout the Pacific are important buildings. They are not just a place of worship but also a place for social gatherings.

Imagine you are the church in the picture. Tell a story about an event eg the day the missionaries began building you.

RESOURCE 2 — Tapu, Manu, Custom

Tapu refers to forbidden or evil. A place or person can be tapu. Mana refers to spiritual power or standing. Cannibalism in the Pacific was linked to mana. By eating an enemy you would get that person's mana or strength. Custom refers to traditional life. It is linked to the extended family and village, and with sharing. When Europeans first came into the Pacific, they did not understand Pacific customs such as communal ownership of land. This led to conflict between the two groups of people.

Create a chart to show the 3 important Pacific ideas.

RESOURCE 3 — Head-hunting and Cannibalism

Early Europeans in the Pacific helped spread the idea of islands as places of tribal wars, head-hunting and cannibalism. The Solomon Islands and Papua New Guinea were known as places to avoid being shipwrecked on in case you ended up in the cooking pot. Solomon Islanders used human heads in many ceremonies such as the launching of new war canoes. Raids took place between islands to get heads. Cannibalism and head-hunting have disappeared.

Talk about why some people get annoyed by the interest of other cultures in the Pacific's past reputation for head-hunting and cannibalism.

RESOURCE 4 — Tikopia

Tikopia Island is in the Solomons. Four chiefs ruled Tikopia. The chiefs held court in their huts. Those wishing to see a chief had to crawl on all fours. Then they touched noses to knees. Only when the chief noticed them were they allowed to lift their heads. To leave the chief, they had to crawl away backwards.

   If you visit Tikopia, you can still see people crawling to a chief.

Decide if you would like to be a Tikopian chief. Share your reasons with your group.

RESOURCE 5   Changes

Samoa is an example of how traditional society is changing. Fa'a Samoa means the Samoan Way. It is about how Samoans are meant to behave. For example, you must always respect those older than you – matais (chiefs), ministers, politicians, doctors, teachers. Fa'a Samoa has two legal systems. The western style system is run by police and the Justice Department. The traditional system is run by villages. Samoan families are usually large. Traditionally family members, along with the aiga (extended family) worked land that the chief allocated to them. Many families today encourage children to work in town for money. And more Samoans live outside Samoa than in it. Most send money back home.

> 1 Explain Fa'a Samoa.
>
> 2 Make a diagram to show what an extended family is.
>
> 3 Outline the difference between communal and individual ownership of land.

RESOURCE 6   The Place of Women

Pacific women are slowly finding their voice. But they still have the second lowest parliamentary representation in the world; they are only slightly higher than that of Islamic-Arabic women. In 2006 Solomon Islands had no women in parliament. Papua New Guinea had 1 out of 109 MPs. Fiji had 8 out of 71. Samoa had 4 out of 40. Niue had 3 out of 20.

> Think of reasons to explain the low figures for female MPs.

RESOURCE 7   Issues Not Spoken Of

Papua New Guinea, for example, has an HIV/Aids epidemic. Females are frightened of gang-rape by packs of 'raskils' or even police. Health workers and teachers tell of parents so desperate to get money for school fees that they send daughters to work the streets. People are slow to talk about these issues and discuss solutions.

> In your group, suggest reasons that most violence against women goes unreported. Suggest ideas for encouraging people to discuss issues.

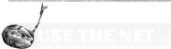

Choose one thing from the unit to research eg Tikopia or mana or head-hunting.

Examples of Traditions

 ## RESOURCE 1   Land Diving in Vanuatu

The people of Vanuatu are believed to be the inventors of land diving. Bungy jumping is based on land diving. Legend says it began when a woman climbed a tree to escape her husband. He followed her. She threw herself to the ground. Out of his mind with sorrow, he followed her. He did not realise she had tied vines around her ankles.

Once the first yams were harvested, the people built a tower high in the trees. Each diver chose his own vine. The aim was for the diver's shoulders to touch the ground and make it fertile for the next crop. The males of Pentecost Island still carry out this tradition.

> Describe how bungy jumping differs from the land diving of Vanuatu.

 ## RESOURCE 2   Stone Fishing in Tahiti

Taha'a island in Tahiti is called The Vanilla Island for its many plantations of spice. It is also famous for continuing the tradition of stone fishing. Villagers wade into the lagoon. They beat the water with stones tied to ropes. This scares fish and drives them ashore. People collect the fish for a feast.

> Explain a tradition in your country that you could take a person from Taha'a to see.

 ## RESOURCE 3   Coming of Age

An example is the haircutting ceremony in Nuie. This is a big event for a teenage boy. The long tail of hair that he has had since childhood is cut off. Guests are invited to a feast. They give money to a fund that goes to the boy after the celebration costs are paid. There is a similar ear piercing ceremony for girls.

> Suggest why these ceremonies are usually held in private homes.

 ## RESOURCE 4   Kava

Kava is a drink. It is part of formal ceremonies and an ancient tradition. Polynesians found kava cured stress and sent the user to a higher plane of consciousness. They also served it to guests and visitors as a sign of goodwill. Kava is made from roots of the pepper plant. The traditional way is to clean the roots, chew them up and spit them into a bowl of coconut milk and water. Then you strain and keep squeezing juice out of the roots until the juice runs clear. The guest of honour is invited to drink first. You should gulp it down in one mouthful.

Kava ceremony for visitors to Fiji.

> Europeans, especially missionaries, tried but failed to stop the use of kava. Suggest reasons why they wanted it stopped and why they failed.

RESOURCE 5   Fire

The men of Beqa's Sawau tribe in Fiji can walk on hot stones without being hurt. An ancient legend tells how an ancestor of the tribe got this power. He was fishing for eels in a mountain stream and pulled out a spirit god. The god pleaded for his life. He said he would give the man power to walk on hot stones. So the man let him go free.

When the king of Tonga died in 2006 people lit fires of coconut fibre around the palace so there was light for the king. This tradition is known as the takipo.

> Explain why such traditions are best left for the experts.

RESOURCE 6   Names

Checking out lists of names of competitors at sporting competitions gives a good idea of what names from different countries are like. At the 2006 Commonwealth Games held in Melbourne in the team from Kiribati were Toom Annaua, Ieie Matang, Rabangaki Nawai, Mariuti Uan, Tierata Taukaban, Tarieta Ruata, Tokannata Ioatene, Toaaki Taoroba, Allie Johnny, Nooa Takooa, Atea Tetabo, Selima Tetau, Taraia Mwaitonga, Tekaei Temake, Beru Karianako, David Katoatau, Teataua Tiito, Marea Taomati, Kokoria Iabeta, Rokete Tokanang.

Among others at the 2005 South Pacific Mini Games at Palau were Esther Meallet of New Caledonia, Iliesa Namosimalua from Fiji, Moi Koime from Papua New Guinea, Aukuitino Hoatau from Wallis & Futuna.

> 1   In 2 sentences write your personal response to these names eg I have heard few of these names before …
>
> 2   In 2 sentences write how important names are to the idea of citizenship eg passport.

USE THE NET

Caroline Mytinger was a skilled portrait artist. In 1926-30 she and a friend went to New Guinea and the Solomon Islands. Most of the $400 they carried was set aside to be used to ship their bodies home, if needed. It took a year to get to the South Pacific by steamship. They spent the fourth year of their journey recovering from malaria. Caroline's paintings are time capsules of lost tradition. New Zealand anthropologist, Raymond Firth, lived on Tikopia in 1928 and 1929.

Choose either Caroline or Raymond and prepare a fact sheet about their time in the Pacific.

Food

 RESOURCE 1 **Importance of Food**

Food represents wealth, generosity, and community spirit. Visitors are usually asked to share a meal. Food is often given as a gift.

Some villagers cook on outdoor fires or kerosene stoves. They cook a traditional feast in an underground oven such as the umu in Polynesia and lovo in Fiji. Many places have modern cooking facilities.

Atolls have poor soil, a shortage of fresh water, and salt spray. Other islands have better soils, water supplies and a range of climates. They can grow more varied food. Many islanders live by subsistence farming. Their crops include sweet potato, banana, maize, cassava, taro, yam, breadfruit, papaya, beans, avocado, mango, plaintain, noni. Some islands raise pigs or beef for locals. Cash crops grown for export include coffee, cocoa, coconuts, palm kernels, ginger, black pepper, sugarcane and tea.

1 Find evidence that food is an important part of island culture.

2 List advantages and disadvantages of underground cooking.

3 From the following, pick the one food that is most likely grown for use only by the villagers: cocoa, sugarcane, tea, yam, ginger. Explain how you found the answer.

 RESOURCE 2 **The Coconut Palm**

The coconut palm is called the most useful plant around the Pacific because it provides everything to sustain life. It can live as long as 100 years. It can produce 50 -100 coconuts a year. Copra (meat) is eaten raw or used in cooking. After oil is pressed out, the meat can be used for cattle-feed. Coconut milk is made from grated pulp mixed with hot water. Coconut water is drunk straight from the coconut. Sap from unopened flower clusters is used to make sugar, vinegar and an alcoholic drink called arrack. The unopened shoot at the top of the tree can be put in salads.

In a group present a mime on how to open a coconut.

 RESOURCE 3 **Legends**

Legend 1

The father of a young boy went back to his homeland. The boy longed to meet his father. He asked his mother to help. His mother chanted to their ancestor, the coconut tree. A coconut sprouted in front of her. She told her son to climb the tree while she kept chanting. The tree stretched and grew over the ocean. It reached the father's homeland. The boy met his father.

Legend 2

An eel lover told his sweetheart that jealous rivals would kill him. You must then cut off my head and plant it, he said. From my head will grow a tree with a fruit which will give you both meat and drink. On the fruit you will see the 2 eyes that adored you and the mouth that spoke love words to you. When the lover was killed, his sweetheart planted his head. From it grew the coconut palm.

1   Explain the meaning of the word 'legend'.

2   Work out why there would be different versions of these legends in the Pacific.

3   Choose one of the legends and turn it into a comic strip.

 RESOURCE 4   **Feasts**

Feasts are a tradition on many islands. The luau, for example, is known as Hawaiian barbecue. It is a big party where the main course is a whole kalua pig steamed in an umu. At the feast you might also eat pupus (Hawaiian appetisers such as spare ribs), laulaus (meat or fish steamed in taro leaves), limu (seaweed), lomi lomi (salmon cooked with tomato and onion), and haupia (coconut and pineapple pudding). A Tongan feast may have 30 different dishes such as steamed pork, suckling pig, fish, crayfish, beef, octopus, and tropical fruits spread out on a long tray called a pola.

1   Describe how a feast you have attended compared to a luau.

2   Explain why feasting is such an important way of keeping culture alive.

RESOURCE 5   **Western Foods**

Pacific Islanders used to have no <u>access</u> to Western foods. They ate fresh fruits, vegetables, poultry and seafood. Now they can get highly <u>processed</u> foods such as white flour, white sugar, white rice, canned meat and fish, margarine, <u>mayonnaise</u>, biscuits, soft drinks and fast food. These foods are low in <u>fibre</u> and high in fats and sugars. Some islanders buy these foods to eat, and sell their fruits and vegetables.

The change in diet is one reason that diseases such as heart disease, <u>hypertension</u>, type 2 diabetes, and <u>obesity</u> are on the rise.

1   Describe the difference between traditional foods and highly processed foods.

2   People often find it hard to spell the underlined words. Learn the spellings. Then write or draw a summary of the resource using all the underlined words.

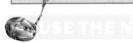

1   Besides food, the coconut has many other uses. List them.

2   Collect a recipe for a dish, featuring traditional food, that you would like to eat.

UNIT 28 Craft

RESOURCE 1 — Regions

Each region has its own special crafts. In Vanuatu, for example, traditional crafts include necklaces and ankle rattles made from shells. Master carvers carve bowls and plates, masks, headdresses, tree fern statues, clubs and spears, bows and arrows, miniatures such as canoes and animals. They often decorate masks with paint, feathers, or pig tusks. Some islands have ancient rock paintings and caves decorated with paintings of animals. People on some islands use sand drawings to tell stories.

In many Pacific regions, pandanus and coconut palms are slit into ribbons for weaving baskets, fans, hats, mats. Hats are often decorated with small shells which are painted and stitched on by hand. The women of the Cook Islands are famous for handstitched quilts called tivaevae.

1   Work out what the word 'craft' means. Give 2 examples of a craft.

2   Explain the link between craft and culture.

3   Explain the link between craft and heritage (objects passed down generations).

RESOURCE 2 — Tapa

Tapa is the traditional cloth found on islands such as Fiji, Samoa, Tonga, Papua New Guinea, the Cooks. It is used to make clothes, mats, wall hangings, linings for baskets. At events such as weddings and funerals large amounts of tapa cloth are used as gifts.

Tapa is made from the inner bark of the mulberry tree. Bark is stripped, soaked in water, scraped with shells, pounded into thin sheets by wooden mallets. When you visit one of the tapa-making islands, a constant sound is the mallets beating out lengths of mulberry bark. Then sheets are pressed together, and dried in the sun. The cloth is light and felt-like.

Women stencil red, black and brown traditional designs on it. They use natural dyes made from bark, seeds, and clay.

1   Be a tapa maker in Tonga. Give a reason why your group is a source of news and information.

2   Draw a mallet and explain its importance in Pacific culture.

3   Create a set of diagrams to show how tapa is made.

RESOURCE 3   Dance

The traditional Hawaiian hula is a flowing dance that tells a story. It began in ancient Hawaii as a form of worship. It can be done with or without music.

Musical instruments are made of gourds, coconuts, or logs covered with shark-skin membrane. The ukelele is popular. Dances called mekes in Fiji tell stories such as how a war was won. Some are passed down through generations. Others are specially created for the new event. The seasea is a women's graceful fan dance. A cibi or meke wesi is a men's fierce spear dance. A traditional Tongan dance is the lakalaka. Each dance tells a new story. Many Samoan dances are about everyday activities.

1 Explain the link between dance and events.

2 Explain the link between dance and culture.

3 Explain the link between dance and heritage.

RESOURCE 4   Whale's tooth in Fiji

A whale's tooth (tabua) is a prized gift in Fiji. It is used only for very important presentations, or when someone is making a request or when someone is saying sorry.

Suggest a reason the tabua is so prized.
Think of an equivalent gift in your culture.

RESOURCE 5   Tattooing

Tattooing was once common. Missionaries tried to stop it. But some societies kept it alive. Samoa and the Marquesas, for example, are famous for tattooing. In Samoa, the tradition of tattoo, or tatau, done by hand by tattoo artists called tufuga was often passed from father to son. The trainee spent hours tapping designs into sand or tapa cloth using a special tattooing comb, or au. Ceremonies were held to tattoo young chiefs. The tattooing and healing took many months. The tattoo showed their bravery and how they honoured their culture. The traditional pe'a covered the body from mid-torso to the knees. In many pe'a, a boat sat on top. It stood for the voyage to Samoa that the first people made.

Women had smaller tattoos, most often on the thighs, legs or hands. Tattoos on the hands, called lima, were needed to be able to serve kava.

1 Talk about why missionaries tried to stop tattooing, why they often failed, and why an unfinished tattoo was a mark of shame.

2 Talk about the place of tattoos in culture.

USE THE NET

1 Find out the importance of flowers to Pacific cultures eg leis stand for friendship or love.

2 Prepare a list of traditional crafts in one Pacific country.

UNIT 29 Sport

RESOURCE 1   Kilikiti (aka kirikiti)

Missionaries introduced cricket to places like Samoa. From it came kilikiti. The bat has 3 sides. This makes controlling direction hard. An average bat is 1.2m long. It is a cross between a baseball bat and a war club. It is sennit-wrapped, wooden and painted. The ball is made of hard rubber wrapped in pandanus. Players are not protected by pads or masks. Often they wear only lavalavas. Rules are flexible. There are no limits on the

number of players in a team. Teams have males and females of all ages. Kilikiti is played all year round. Traditionally a game took up to 3 days to finish with whole villages taking part. Singing, dancing and feasting were important to the game. Since 1999, NZ has hosted the international Supercific Kilikiti Tournament.

1 Draw and label a lavalava, kilikiti bat, kilikiti ball.	2 Write a few sentences comparing kilikiti to cricket.

RESOURCE 2   South Pacific Mini Games 2009

Host = Cook Islands. Length = 10 days. Countries = 22. Athletes = about 2,500. Sports = 15. Budget = NZ$15 million. New buildings = 3 vaka sports centres, 2 outer island sports centres. Past hosts = Solomon Islands (Honiara) 1981, Cook Islands (Rarotonga)1985, Tonga (Nukualofa) 1989, Vanuatu (Port Vila) 1993, American Samoa (Pago Pago) 1997, Norfolk Island (Kingston) 2001, Palau (Koror) 2005.

1 Work out what following figures refer to: 10, 4, 7.

2 Decide what the places in brackets most likely have in common.

3 Suggest 3 benefits (good things) that could come to the Cook Islands for hosting the Games.

RESOURCE 3   Water Sports

In your group talk about how these sports help make the Pacific popular with tourists.

USE THE NET

Get the names of members of any 2 national sports team from your country eg netball and rugby sevens. Find out the country of origin for each member. Highlight any from Pacific Islands (excluding your own country).